TOR FALCON

Sugar Beet MOON

An Artist's Study of the Sky

With texts by Andrew Lambirth & Dr Dan Self

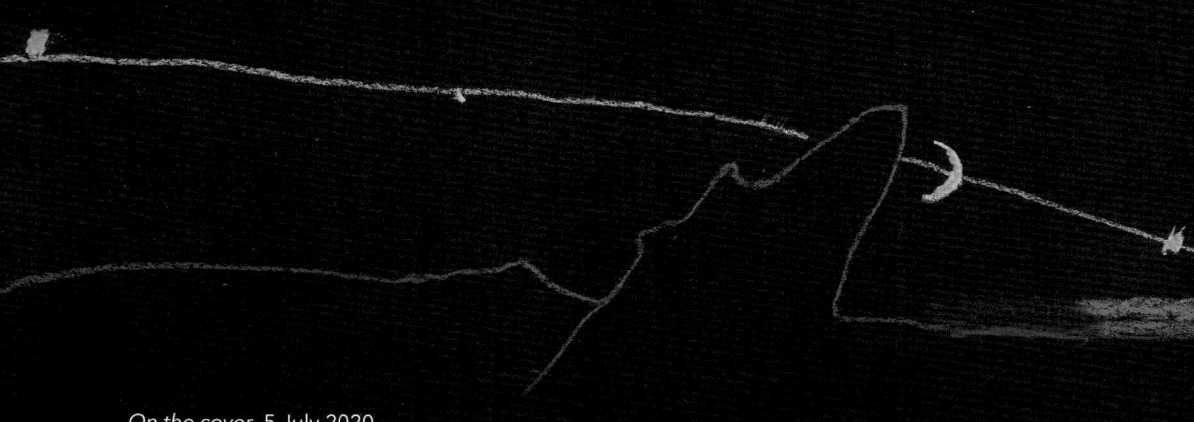

On the cover 5 July 2020
above Ecliptic, December 2020
pp. 10–11 21 September 2020
pp. 48–9 Sunset, 25 January 2021
pp. 124–5 Moon phases
p. 133 Perseid meteor shower, 12 August 2020
pp. 134–5 18 February 2021
p. 136 Photo by Harry Cory Wright

p. 4 Extract from Robert Frost, 'The Death of the Hired Man'
p. 5 Extract from Thomas Gray, 'Elegy Written in a Country Churchyard'
p. 13 Wendell Berry, 'To Know the Dark' (published in *The Peace of Wild Things*, Penguin, 2018)

In order to protect their anonymity, the names of some individuals in this book have been changed

Designed by Niki Medlik at **studio medlikova**

ISBN 978-1-3999-8791-2

A CIP record for this book is available from the British Library

Printed and bound by Swallowtail Print in Norfolk

CONTENTS

FOREWORD

TOR FALCON: MOON WATCHING

Andrew Lambirth

Part of a moon was falling down the west,
Dragging the whole sky with it to the hills.

Robert Frost

The nocturnal world has consistently fascinated artists. Painters as varied as Whistler, Van Gogh, Atkinson Grimshaw, Emil Nolde, Caspar David Friedrich and Edvard Munch have invoked the moon in their work to memorable effect. Closer to home, the Romantic tradition of Turner, Blake, Samuel Palmer and Paul Nash has sought out the moon as symbol and guide. The dramatic crash of colour at sunset foreshadows the descent of the world into darkness, before the moon rises and casts its pale glamour over night operations: hunting, love-making, troop manoeuvres, sea voyages. The moon as pilot is but one role among many.

The moon is traditionally associated with fertility and regeneration, resurrection and immortality, occult power, mutability and intuition, and the emotions generally. Despite this emphasis on change, it is also a regulator of time: the cycle of the tides and seasons, crop growth and the lives of women. In itself it makes visible and actual the great cycle of birth, growth, death and rebirth. Today we regulate our lives on the Gregorian calendar, which, although a solar calendar system, evolved out of the lunar calendar – the earliest way of measuring time. The moon remains central to all our lives, as poetic symbol and actual reference point.

Although some cultures identify a male moon god, the moon is usually more closely allied with the female principle, though even in England we speak of 'the man in the moon' – the imagined male face appearing in the great sphere of the full moon. Sexually ambivalent, the moon appeals to all. Alongside scores of other artists, Tor Falcon has been seduced by the lyric moon and the liminal world of dusk. It has been said that daylight reveals, and moonlight mystifies, but nothing is quite so simple. (Think of the 'Bomber's Moon' that reveals everything.) Each moon is the same and yet different. Falcon says: 'To watch a moon rise makes you feel euphoric.' It makes her laugh; for her it is a happy, joyous thing to be doing. In her latest work, she wants to achieve on paper some equivalent of the moon's singular beauty and share it with us.

Tor Falcon's previous projects have included walking the Peddars Way from Knettishall Heath to Holme-next-the-Sea, and exploring the rivers of Norfolk. In each case a significant body of work has emerged from her investigative peregrinations, the drawings being made along the way in front of the subject. There is a tradition of solitary wanderings in the literature and art of these islands that stretches back to Old English poetry. Falcon brings the peripatetic impulse into line with contemporary issues, writing a commentary on her discoveries (the people she meets as much as the landscape she encounters), which is more picaresque journal than artist's pattern book.

With her new theme, the moon, she has allowed herself to be pinned down to one place: the garden and surrounding fields of her East Anglian home. This is an enveloping countryside of woods and meadows, without commanding vistas or high viewpoints. She draws the moon while remaining firmly grounded near her studio. This could be oppressive, especially with melancholy night closing in, but it isn't. Darkness promotes anonymity, and can be frightening and isolating. Instead, the landscape here is sympathetic and resonant, full of mystery but not of harm. As the gloaming advances, it might be the setting for Thomas Gray's 'Elegy Written in a Country Churchyard':

> *Now fades the glimmering landscape on the sight,*
> *And all the air a solemn stillness holds.*

It is as if the earth and air are waiting for something, and of course what they're waiting for is moonrise.

The original idea behind these three series of works is that each chalk drawing be completed on the spot in one sitting, and thus become the most direct kind of record that can be achieved. Of course, there is some leeway in this mode of work, and some drawings have to be completed or re-worked in the studio. Sometimes, entirely new works are made from the on-site studies, usually larger and more elaborate. Falcon's earlier drawings, such as those made in her year-long survey of the Peddars Way, tended towards flat designs on the page, all intriguing and often intricate surface pattern. The latest work has deepened the game and advanced the spatial enquiry, and the forms are much more three-dimensional. In fact, the tendency now is towards the sculptural. Perhaps moonlight, and the deep shadows it creates, accentuates the sculptural presence of things. Certainly her work has grown richer for this increasing solidity.

The work in her first two books could be called travelling drawings. She wrote in *Rivers of Norfolk*: 'I have looked intensely. I have used drawing as a means of exploring.' Her moon work is less about movement through a landscape and more about being the fixed point in a turning world. She is still looking intensely, never more so, and she is still exploring, but for this project she remains rooted in her own parish. The work's dynamic is thus necessarily different. Falcon observes that the moon appears to change in every place you see it, that its surroundings to some extent dictate the way you see it, as well as its mood (or the mood of the viewer). She says the moon in her garden, in her fields, is a bit of a trickster, disappearing behind trees, reflected in water. 'I really love the chase,' she says of her pursuit of the image she wants to draw.

When she started investigating the moon as a subject to paint, Falcon conducted considerable research into the nature of this satellite planet and its habits of motion through our skies. It is easy to be overwhelmed by scientific data once you start asking the whys, the wherefores and the whens, so she began to draw it: firstly the classic full moon, and then continuing to concentrate on it when it was rising. Although it might seem simple and straightforward to start drawing, this was actually a lengthy process, partly because the look-out point had to be right. In fact, it took her a year of experimenting to find her ideal landscape of moon watching. Although a traditional place to watch the moon is from a roof or tower, she decided she preferred to be grounded rather than raised on a platform.

Night drawings are in a sense the reverse of daylight pictures: the landscape is effectively dead ground and the sky is where things are really going on. But even something as seemingly self-evident as that took time to come to terms with. When she first attempted to draw the subject, Falcon persisted in thinking the picture was still all about the foreground, not the background sky. Drawing the moon also made her more intensely aware of the activity of looking. 'I got very interested in how my eyes worked,' she says, 'and what appeared to me as the light got lower, until the last bit of colour had gone.'

As it gets darker, am I seeing colour or imagining it? This is a question she constantly asks herself. Or indeed am I simply remembering it? For this is territory she knows well. Once again she kept a journal, and wrote it up immediately after a moon watch. As she grew to know her subject, she realised it was not necessary to understand the moon in a scientific way. It was enough to marvel and not fully comprehend: she speaks of this as 'a joy'.

Before going out for a drawing session she would tape small pieces of paper to a drawing board, so that she could make out their edges in the dark. (Paradoxically, she sometimes uses black paper, which is almost invisible. She also makes good use of pinky buff sugar paper.) She would then select chalks with which to draw, and made sure she knew which was which, mostly identifying them by touch (labels, size, and so on). Drawing in the dark must be rather like drawing and deliberately not looking at what you're doing, or drawing with your eyes closed, both accepted studio exercises for artists. In some respects, being there and making marks on paper – just the act of witnessing and recording the drama of the skies in this way – is the essence of the experience. At this point it's not about making an attractive picture; a few lines are enough to connect with the natural magic of the moment.

As regards her choice of chalks, greens and greys have always been her preferred colours, and as she says, 'I had a manageable little palette which pretty much worked everywhere. When I started drawing in the dark, a rainbow of colours opened up. I was surprised by how much colour there is in the dark. I bought loads of new pastels (probably too many) – colours I'd never have looked at before, like arsenic green, vermillion, etc. They only need using sparingly, though. I'd say pink is the colour I use lots now. Pinks, reds, oranges and darks.' In addition to her distinctive use of colour, Falcon employs a soft conté pencil to draw into her forms or add definition in terms of outlines.

She does not work in a technical, astronomical kind of way, but she is interested in charts of position and movement, which she draws out from time to time, to help her understand the music of the spheres. *Musica universalis* is an ancient philosophical notion that the mathematical relationships or proportions in the movements of the planets create a kind of music. Pythagoras believed this, and his idea was taken up by the 16th-century astronomer Johannes Kepler. As a poetic metaphor it has lost none of its power.

As she draws the gradual ascent of the moon, the sun is behind her, setting: one could not exist without the other, and each defines each. They are the opposites or dualities of our existence. So of course the sun is just as much a part of this process, and realising this propelled Falcon into a year of drawing sunsets, one a month. She loves the challenge of her subject: 'I'm really interested in drawing things that I can't draw.' And how to convey moonlight remains a problem. The moon is coloured, but moonlight is not. Some of her most beautiful studies are of the gloaming, and particularly the sunset's red afterglow, which lasts for about five or ten minutes and which Falcon finds 'compelling'.

The new work demonstrates her greater authority with her chosen medium of chalk or pastel. This is manifest in two ways: through a greater control (less scribble), and paradoxically also through a greater looseness (the ability to make the slightest line or squiggle stand for a whole nexus of things and events). Her forms are more crisply outlined and more thoroughly realised, making reference in their general approach (rather than in specific examples) to the landscapes of John Nash, an artist she much admires. Although she works in an accepted realist tradition, she brings her own interpretation to it, and her work is so well-observed that it remains fresh to the eye and mind.

In many ways, the drawings made on the spot are the real statements, not the ones composed later in her studio, however seductive and intriguing. For instance, a series of four tall thin drawings, done on wallpaper, celebrate and record the month of October, describing the position of the moon in each quarter of the month. These studio drawings have a completely different character from the *plein-air* work, but if they have none of its spontaneity and immediacy, they do allow Falcon to explore other qualities and formal possibilities discovered in the landscape. They are beautiful works in their own right, but they don't have the sense of risk that work outdoors can offer. On one occasion when she was drawing in the fields, a fireball suddenly appeared, a meteorite burning across the heavens. She wouldn't have seen that – or been able to draw it as it passed – in the studio.

The moon, and the sun likewise, are what Coleridge called 'obstinate in resurrection'. Thank goodness they are. We cannot live in a world in which there is no moon, or the sun never rises. Their cycles are our cycles, their light our light, their visual dramas a great and sustaining nourishment that so many of us take for granted. Living in cities it is often easy to ignore the sun and moon and think they do not matter. And how many times have you travelled on a commuter train when most of the passengers seem entirely oblivious to the glories of the sunset taking place outside the speeding windows?

Tor Falcon pays close attention to what the rest of us may miss. Her way of coming to know the world is through her chalks, wielded with an alert eye and an enquiring mind. As she wrote in *Peddars Way*: 'Drawing is a way of stepping out of one's busy life, of just being a human in a landscape, of sensing your size in relation to what is around you, but mainly, of just taking time.' She reminds us to take time: firstly to look at her drawings, and then to ponder our relationship to what she has drawn. Thus does she share her enthusiasm for the glories of the world around us, and enhance our understanding of it.

opposite 28 March 2021

YEAR 1
JANUARY 2020–21

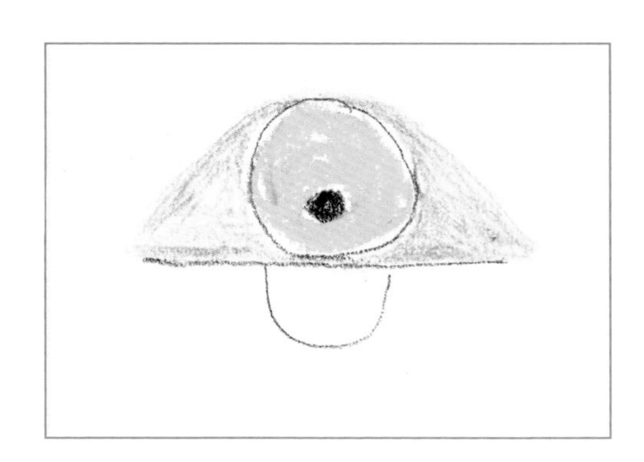

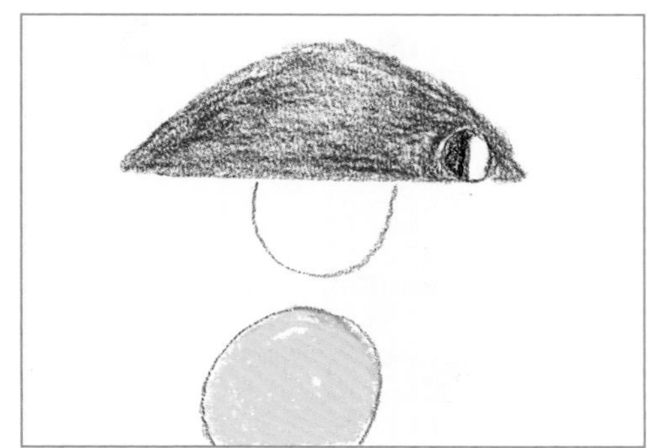

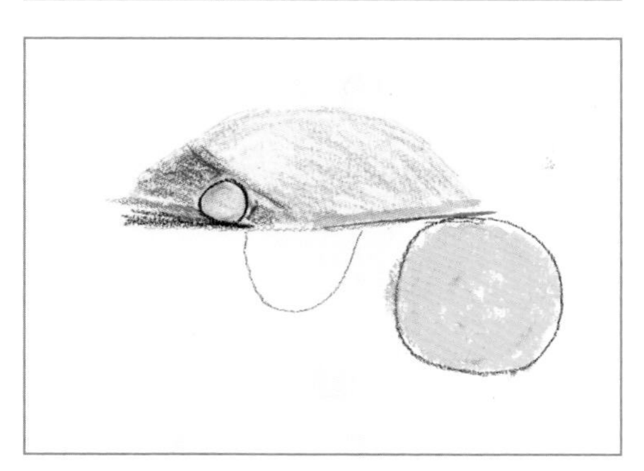

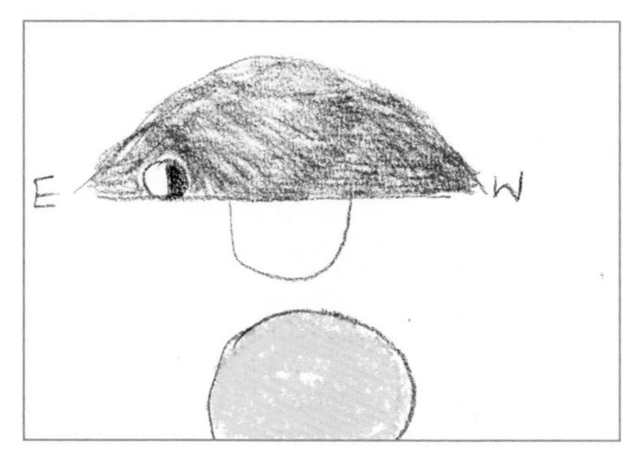
E
W

YEAR 1

This started slowly. And nearly ended immediately. A mild curiosity about the timing of full moon rise led me to my neighbour's field one January afternoon. I wanted to watch and draw the full moon rise. Drawing is my way of really paying attention. You've got no choice if you want to coordinate your brain, eye and hand. I wanted to know if the full moon rose at sunset throughout the year. At 15:30 in December, at 21:30 in June, and everything in between? I hoped that by going out and drawing the full moon rise I could settle this question and perhaps even learn something about the moon. I knew there was a lot I didn't understand. What started as an idle once-a-month bit of inquisitive drawing led me gradually to what felt like a new world – the 'blooming and singing' dark world of a poem by a radical sage. My narrow full moon rise remit expanded and within a few months became a preoccupation that encompassed everything around me.

> To go in the dark with a light is to know the light.
> To know the dark, go dark. Go without sight,
> and find that the dark, too, blooms and sings,
> and is traveled by dark feet and dark wings.
>
> Wendell Berry

opposite Drawings visualising the four stages of the moon, in relation to the sun and the flat horizon, showing both night and day

10 JAN

Setting off to John's meadow. 15:30. Clear, still, cold. Hoping to draw the full moon rise, which I think should happen around sunset, and I reckon I'll get quite a good view looking down the stream: that's roughly E. I've noticed that the full moon seems to rise at sunset – well, it has since October when I first started looking. Although last month it was pouring with rain, so I can't be sure when the full moon rose then. I'm curious to know what happens in June, when the sun doesn't set until 22:00. Perhaps everyone knows? Is it common knowledge? Why don't I know? I'm going to try and make a point of going out and drawing it – as a way of really concentrating on looking. Hopefully I'll gain some understanding.

My back is to the setting sun and its light is turning the meadow and trees blisteringly pink. 5 noisy Canada geese fly E, their bellies and underwings the same colour as the vegetation. I can't see the moon, so I try and draw this fluorescent light – which seems to be intensifying. It really is pink. I look down at my paper and look back up, and suddenly there's the moon, in the gap over the stream. Bloody hell! My scalp prickles. I wasn't expecting it to be so big. Or perfect. Or pale. An enormous, slightly translucent, milky circle that wasn't there a minute ago. I'm shocked. I stop drawing, I can't take my eyes off it. I think my mouth is probably agape. It seems to be turning pink. Is it? I stare and stare and stare, as if this is the first time I've ever seen the moon. I remember I want to draw it and realise the foreground has changed colour. The pink has faded and is turning dull purple. Grab another bit of paper. Crap drawing, the moon has moved. The trees are dark silhouettes now. Another bit of paper, another rubbish drawing, all the colours have changed. Another bit of paper, another attempt. The sky is yellow, no it's pink, no it's mauve. The foreground is getting darker and darker. The moon is moving up and S. The moon is changing colour, it's golden pink now. It's getting dark, I can't work out which colour pastel I'm picking up. Just above the horizon there's a strip of dark grey purple sky. Is the grey spreading up? It appears to be chasing the moon, which is orangey gold now. I suddenly notice the moon's reflection in the cattle drink in the stream, in front of me. Darker than the moon in the sky, broken and squirming in the water. I can't see my pastels or my paper. A moorhen starts screeching. I've stopped drawing, I have to, I can't see anything, but I go on staring. It's so dark.

That was impossible – an hour and a half, moving subject matter, fading light, colour chaos, 5 terrible drawings. Where are they? With a sinking feeling I realise my daytime method of strewing pastels, paper and pencils all over the place doesn't work in the dark.

Spend ages feeling around for discarded chalks. The moon is now above Mr and Mrs Stagg's house at the top of the hill and is silvery in a more-or-less uniformly dark sky. A heron shrieks.

I walk home simultaneously elated and crestfallen. I'm buzzing, amazed and gobsmacked by the moon, but dismayed, embarrassed and depressed by my drawings. I'm irritated that it got dark. Back in my studio I briefly want to cry. I have managed to turn the most exquisite thing I've ever seen into these awful drawings. It's frustrating having to stop after one attempt. If I was drawing something difficult in daylight, I'd try again and again and again. Darkness is so final and I can't rewind the moonrise. I feel impotent. I still feel wildly elated. I'll go tomorrow with a torch.

11 JAN

15:30 Another clear day. John's meadow again. I have a torch. I draw with my back to the sunset. The field and the trees are that hot pink again; I've only got two pinks and neither of them are right. I do another drawing with my back to where the sun has just gone down. I can't decide if the sky is coloured at all. Is it slightly yellow or very faintly blue? No moon. Behind me the sky is throbbing, it's orange, red, yellow and purple – I feel I'm looking the wrong way. It gets darker and darker. The eastern sky is now a gradation of pale pinks and purples.

above 10 January 2020

Then the grey shadow appears on the horizon again and moves silently heavenward obliterating colour. No moon. Twiddle my thumbs. I do a completely dark, completely pointless drawing. Ducks flying about. No moon. I stand up. Is that a faint glow I can see above the stream? I wander about. No, no glow. Feel stupid. It gets even darker. Where the hell is the moon? 17:00 Am I imagining another glow, behind the big ash tree? It really is a glow, it's getting stronger, pinker. More powerful. Yes…finally a glimpse of pink orange. Heart-stopping. Pure colour. The moon. At last. And rising halfway up the hill. What the hell? It's a dreadful composition from here. Why is it up there? Why has it risen in a different place, and why is it so much later tonight? Drawing in the dark is impossible and using a torch isn't that great. Turn it on, turn it off. Can't seem to see colour in torch light. I'm alternating between being dazzled and then blinking blindly in the dark. God, this is difficult. I need another hand. At least I can see to pack up, though. I watch the moon climb higher and turn from pink orange to gold to silver. Why? Again I feel elated and furious in equal measure. Really, really pissed off with the dark. Why did I think this was a good idea? Green plover call from overhead. I'm cold. Drawings go in the bin.

12 JAN

16:30 Clear, still, cold. It's getting dark when I get to John's meadow. Tonight I've got a head torch on so I don't have to fumble about turning things on and off. Moon doesn't show. I'm sitting in the meadow, in the dark, doing nothing, feeling ridiculous, getting cold. 17:30 No moon, colder. 17:45 No moon, very cold. 18:00 No moon, extremely cold. Bewildered. Getting cross. Do star jumps. Think about going home. 18:15 Had enough, packing up. Then notice, near the top of the hill, the horizon is beginning to glow. Is that the glow from Norwich? No, it's becoming more and more intense. Real suspense as it heightens. It's almost unbearable to watch. Finally, I can see the top edge of the moon and it's on fire. Wow. Luminous red gold beauty with a crisp edge. In a world that has gone fuzzy with darkness, that clean curved edge of molten colour is thrilling. All cold and annoyance now forgotten. I am jubilant. This – this birth of colour, this globe of light – is something worth waiting for. 18:30 by now, and I can see the moon complete. Its leading edge is less defined. It isn't a perfect sphere tonight. It's to the left of Mr and Mrs Stagg's house. A Chinese lantern behind black paper-cut trees. At the top of the hill. Every night something different. As it rises higher it glows less red. It turns the sky around it purple. The head torch is as hopeless as the torch. My hands might be free but I can't differentiate colour and when I look up into the darkness my eyes are blind, the moon diminished. The composition from here is awful, a large expanse of dark field sloping up to a line of trees and a house. I'd never draw that in daylight – although it's all black anyway, so who cares? Why does the moon rise in a different place every night? Will it be even later tomorrow? Further along the horizon? Why am I doing this? Tawny owl and heron noises. Elation and irritation. More drawings in the bin.

13 JAN | WANING GIBBOUS

Heavy rain.

A waxing moon is a growing moon. A waning moon is a shrinking moon. A gibbous moon is over half full. A crescent moon is over half empty. The cycle goes: new moon, waxing crescent, first (or half) quarter, waxing gibbous, full moon, waning gibbous, third (or half; also known as last) quarter, waning crescent, new moon.

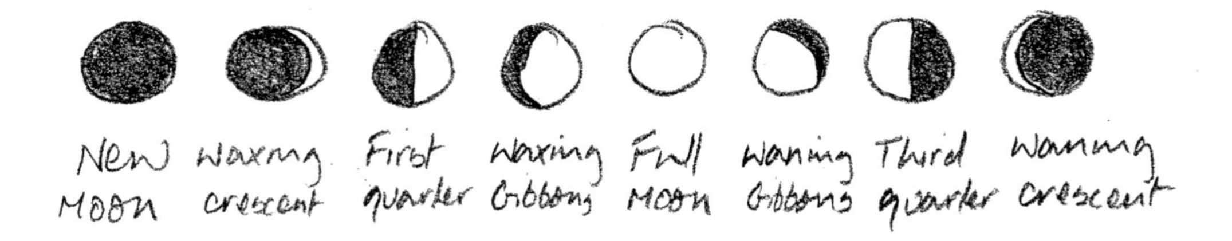

above Moon phases (Northern Hemisphere)

9 FEB | **FULL MOON**
17:15 Clouds, windy, cold. Back in John's meadow, hardly sit down before the moon appears halfway up the hill. Bloody hell. Not above the stream like last month. This is only a good place to draw from IF the moon rises exactly over the stream. I'd been hungrily anticipating its reflection in the stream. I've waited a whole month for a second go. Why isn't it rising in the same place? This is not at all like last month, this full moon is pink gold, it has a red hat of cloud above it. Briefly it appears to be rolling up the hill on the tops of the trees. It launches itself into the air just in time to miss a collision with the Staggs' house. Crows and jackdaws and rooks swirling about in the wind. Bad drawings. I'm having real trouble with colour. There is no bit of this moon drawing idea that I've got the hang of. My subject moves – in space and time. My subject matter isn't predictable. Darkness is always creeping up on me. How do I draw in the dark? My palette of colours is wrong. How am I going to do this?

As I walk back through the garden I turn to look at the moon. Snagged in a row of stream-side alder branches, its reflection is bobbing about in the gravelly tail of the swirly whirly pool.

10 FEB | WANING GIBBOUS
07:30 Walking the dogs, caught a glimpse of the moon, now pink, setting behind the wood. Utterly different from the moon of last night. Smaller and with no drama. A demure descent.

24 FEB | NEW MOON
Freezing, clear. It's 18:45 and I haven't seen the moon for over a week now. I want some more attempts at moonrise but it seems to have disappeared. I've been drawing at dusk, though, to try and get better at it. And I've been reading about twilight. Or twilights. Apparently you can break twilight into three stages: Civil, Nautical and Astronomical.

Civil twilight is when the sun is between 0 and 6 degrees below the horizon. Then comes Nautical twilight, when the sun is between 6 and 12 degrees below the horizon, and finally Astronomical twilight, when the sun is between 12 and 18 degrees below the horizon. Night is when the sun is below 18 degrees below the horizon. At dawn the twilights obviously happen in reverse order. I'm intrigued that the seamless dimming of the light can be broken into three. Why would you want to do that? I wonder if I can tell when one turns into another? Surely I would have noticed before? I've found charts with the exact times for today, and I've been outside drawing from just before sunset to see if I can notice any obvious change in light levels.

Sun above the horizon, I can see and draw without impediment.

Civil twilight, 17:23–17:58, is easy to draw in. The sky is light and there is colour in the landscape, birds sing.

Nautical twilight, 17:58–18:38, colour going fast, sky getting darker, robin sings, it's too dark to draw by 18:20. Ducks, in 3's and 4's, flying in all directions. Hearing their wing beats is easier than seeing them now.

Astronomical twilight, looks like night to me, tawny owl hoots.

There's no noticeable jump from one to the next. Twilight tonight is the usual, dependable, gradual dimming.

Civil twilight is a sadly humdrum description for the alchemy of low light. I think the names probably describe functional things like the shutting and opening of city gates just after sunset or before sunrise. Nautical twilight is when the majority of stars become visible, which was so vital for navigation at sea, I guess.

26 FEB | WAXING CRESCENT

Clear. Very cold. 18:00 Civil twilight rolls into Nautical. A perfect sickle moon in the W. Supremely skinny and laser sharp below a very bright star. Attempting to draw it with a blunt lump of chalk pastel is ridiculous. If I stand on the bridge I can see it, darker and shaking, in the weir pool, which has turned a muddy violet colour. A fox is barking in Luther's yard.

29 FEB | WAXING CRESCENT

Nearly half moon in SW, visible briefly between cold rain showers 20:15. Windy. Twilight darker tonight. It's no good. I can't see to draw long before the end of Nautical twilight. However I do it, a torch only makes matters worse. It makes the dark darker. Fox howling, muntjac barking, tawny owl screeching, quad bike roaring. More bad drawings.

2 MAR | FIRST QUARTER

17:30 This morning's rain has cleared. Woodpecker drumming in the wood. Crows, rooks and jackdaws making a noise. Kites circling. The sun is setting in the W. From the bank of the weir pool its light extends the length of the lake, down the bank, and just catches the tops of the most turbulent water. The moon is half full and incredibly high in the S. Seems very small and far away. It's beginning to glow. It's so high it's off my paper. I want to draw it in relation to the place I'm standing in, but a normal rectangle of paper won't do.

23:00 Clear, very cold, starry, the moon is low and gold over the wood. When it was so high earlier this evening, the dividing line between the illuminated half and the invisible half was vertical, pointing to 12 o'clock. Now it's pointing to 10 past 2. When it reaches the horizon, will it be at quarter past 3? It turns in the sky.

7 APR | **FULL MOON**
It's been a bright, clear, warm day. 19:45 I'm walking to the meadow to draw the moon rise and it's already up. I can see it pale gold and to the right of the alders. Bloody hell, I never get this right. Stop where I am by the swirly whirly pool and try and draw before the light goes. Trees, bushes, blackthorn blossom. Blackbird. A sudden flicker of light in the dark stream. Purple sky, pink moon. Bewitching.

Back in the studio my heart sinks as I survey my drawings. They seem to be getting worse with practice. Colours don't read properly. That blue is not a nighttime blue. Composition doesn't work. Everything about them is atrocious. Doubly so because what I witnessed was sublime. This was a stupid idea. It's obviously something I'm never going to be able to do. It's the one-attempt-and-it's-all-over-for-another-month that's crushing. Feel very disheartened. Cheered, though, when I notice I have a moon shadow as I walk back to the house.

20 APR | WANING CRESCENT
Sunny day. 20:00 The tracks in the wood are carpeted in forget-me-nots. They are looking very blue. I haven't seen the moon for ages. Well, it feels like ages. It just disappears, and I'm sure time goes slower when it's not visible.

25 APR | WAXING CRESCENT
21:30 Looking for my first sight of the new moon. It's quite dark. Stars. Disturb pigeons in Luther's yard as I walk to the big field. Oh! There it is, lying on its curved edge, the tiniest splinter resting on top of Mr Girling's Christmas trees. A very bright star above it. Is that the evening star? What is the evening star? A roe deer barks. I can smell blackthorn blossom. Drawing with a torch is hopeless. Feel more blind with it than without it. The light feels like a barrier to seeing.

Shooting star. Wow, a shooting star. Definitely a shooting star in the E as I walk home. Disturb more pigeons.

5 MAY | WAXING GIBBOUS
18:40 We've moved the water buffalo to the far field so I can come into the big field. The moon has just risen above the trees in the SE, the sun is sinking behind me. Alabaster moon in violet sky. Reddening sun in yellow sky. Shadows lengthening and moving. Warblers singing in the brambles. Thrushes in the thorns. Don't know which way to look or what to draw. Want to draw everything.

20:25 I'm by the stream, the moon is just above the thorns. The stream

above 7 May 2020

is invisible, engulfed in foliage and shadow, murmuring quietly to itself. Blackbird singing. The rose gold moon is alone in a pale purple sky, bar one tiny cotton-wool cloud which is drifting slowly W. As it passes under the moon it turns pink, briefly outshines the moon and wafts on its way.

6 MAY | WAXING GIBBOUS

One day before full moon. 19:15 I'm on the big field drawing shadows, looking at thorn blossom in sunlight and shade. Tsunami of birdsong. Moon rising about now but can't pull myself away from the blossom. 20:30 The moon is behind trees on the other side of the lake. The sky is pink. 21:00 From the top of the big field the moon is above the trees. As the sky darkens and the greens of vegetation turn black, the thorn blossom starts packing a punch. I can see to draw.

22:30 Back in my studio I look at my drawings and groan. What colour IS the sky if dark blue looks so wrong?

Roe deer barking. Owl hoots.

7 MAY | **FULL MOON**

20:50 Go back to the place at the top of the big field. I'm looking SE. From January each full moon has risen further to the S. If I was in John's meadow I don't think I'd be able to see it. It would be over my shoulder, a long way to the S of the Staggs' house. Warm. Gnats. Rich evening song of the thrush at the top of the tallest birch to my left. The moon should have risen 10 minutes ago. Draw fading light.

Crows. 21:10 No moon, getting quite dark. 21:15 There it is – a flash. Behind the trees. Deep amber. I'm holding my breath, the suspense is killing me. Slowly, slowly. The sharp, flawlessly curved, top edge. And finally the circle is free of foliage. Humungous. Orange now, staining the dark sky ochre. The blossomed thorns come into focus. They jump forward. Moon appears imprisoned behind thin dark cloud bars. But only for a minute or two. It escapes through the roof and sails on heavenward. The usual mixture of pure, heart-thumping joy (at witnessing the arrival of the May full moon) and utter desolation (at my inept drawing). My mood has gone from fairly run-of-the-mill sort of middling to off-the-chart high. In fact, it is soaring up there with the moon. Turning on a torch to look down at a scrap of paper feels sacrilegious in front of this glorious celestial being. Totally pointless. Why don't I just forget the drawing? Why don't I just watch?

22 MAY | NEW MOON

No moon. Today the moon is invisible. It's an astronomical new moon. When I am next able to see it, it will be a few days old. A waxing crescent. But that's what I've always called new moon. Order some more pastels in what I hope will be moon/nighttime-friendly colours. More pinks, more pales, different blues.

In the Islamic calendar, a month starts when the waxing crescent is first sighted. This, like everything to do with the moon, isn't a set thing. Depending on the angle between the sun and the moon and the observer, it might be seen 2, 3 or 4 days after the astrological new moon. The new month will start at different times depending where in the world you see it from.

2 JUNE | WAXING GIBBOUS

22:20 Three days before the full moon. Warm evening. Lots of moonlight. I draw the reflection of the moon in the lake above the weir. The black of the water seems bright even in contrast to the shining silver lacework of moonlight round the bottom of the reeds. I'm sure I've seen this scene before, beautifully executed, in many a picture and I almost decide not to bother drawing it as I know my attempt will be ham-fisted. I'm getting used to not being able to draw my subject matter. Why do I keep going? Perhaps because it's so beautiful and surprising and difficult. I'm in awe of this delicate light on the reeds tonight and know I'm treating it too tentatively. I somehow have to grab this subject matter by the scruff of the neck and give it a good shake. This is easier said than done, though, as I can't even see my paper. I drop most of my chalks and blind myself by turning on my phone to search for them. I decide to use my hand to feel for them instead.

It's a bad decision, as I sting and cut both palms without finding the lost chalks. What the hell have I cut my hands on?

3 JUNE | WAXING GIBBOUS
08:00 I find four chalks, lying innocently in grass, just where I thought I'd dropped them. Where are the nettles that stung my palms? Where is the thing that cut me? What on earth was I putting my hands on last night? An unsettling thought creeps into my mind. Perhaps darkness isn't just the normal world with the lights off, perhaps it's another world entirely – and I'm going to have to find different ways to operate in it.

22:45 Moonlight so strong I can draw in it. Just wandering around looking at familiar things taking on strange shapes, like the curved end of the hedge below the tall Monterey pine; it looks like a cartoon phantom. The sky is green in the NE. Billy's bedroom light is the only light on in the house. Its feeble glow below the powerful radiation of the low moon brings a lump to my throat. It's a visual description of the fragility of human life and seems to emphasise how vulnerable love can make us. I went out to look at moonlight, trying to draw in it, looking, looking and looking; I wasn't expecting moonlight to bring on a wild rush of protective love for my son. An otter squeaking from the lake. Splashing.

4 JUNE | WAXING GIBBOUS
19:30 After five months (on and off) of unsuccessful moon drawing I've concluded that the day before a full moon is probably the best day of the month for drawing. It rises about an hour before the sun goes down, so it's not too far above the horizon. I have the luxury of being able to see what I'm doing but it's late enough to appreciate all the colour and luminosity changes. I'm feeling quietly confident and on top of my game tonight, and I'm anticipating a fabulous view of the moon rising over the stream. I'm pointing SE (more or less). I'm all ready. I know what time it will rise. I draw an empty sky over a landscape illuminated by the last light of the sun. I've done quite a lot of these recently. Flag irises in golden light. Flag irises in fading light. I don't know what's happened to the moon. I'm sure it was meant to rise at 19:39, give or take ten minutes. Bugger. I get up and walk back twenty paces – and see the moon to my right; it's already quite high. Hidden behind the thorns on the far side of the river,

opposite & above 4 June 2020

as if tiptoeing past me. I'm amazed by how far S it's now rising. And I feel very stupid. The boys laugh at me – how can you lose the moon? Pretty much every time I go out, boys! I'm sure this little wooded valley is the worst moon viewing place in the world. 21:45 Try and draw the moon above the house like last night, but of course the moon isn't in the same place. It won't be above the house for another hour and it will be completely dark by then. Thwarted again. It seems that there are no second chances when drawing the moon. It won't be in the same place with the same light levels for another year.
I thought drawing crab apple blossom in April was quite a pressured, fleeting thing but actually it's a piece of cake by comparison. I had 2 whole weeks of drawing a tree that was reliably in the same place every day. 2 whole weeks of the sun being reliably in the same place at the same time for 15 hours each day. A complete doddle.

5 JUNE | **FULL MOON**
21:01 Warm, clear. I've finally found the compass app on my phone! I spend the afternoon finding the best place to draw tonight's full moon rise at precisely 127 degrees SE. There's no clear view from anywhere in the garden, so Moz and I wade across the stream and crawl under the hedge. It's not perfect because I'm at the bottom of a hill, looking up, but it is only a Norfolk hill and without the blockage of thorn right in front of me I can see a wide horizon. I set my chalks out and wait. And wait. A blackbird is singing above my head in the hedge. Have I got

it wrong again? The sky is the palest, palest non-blue. I wait and I wait. At 21:40 I decide that I've had it with the bloody moon, there are better things I could be doing with my time. I stand to pack up and leave, but I'm stopped in my tracks by the top of an unearthly bald head poking above the horizon a long way to the S. Perhaps I'm hallucinating? An understated, supernatural, polished pink dome has popped up on the other side of the Hingham road. In an instant I'm in love with the moon again. As it rises the sky remains light and, as always, I find it impossible to draw. How do I make the moon brighter without darkening the sky? With each degree climbed above the horizon, the June full moon transforms itself – from subtle pink to showy rose gold to can't-look-anywhere-else tangerine. The top of the hill begins to glow and then, slowly at first, a train of light spills down and across the field towards me in the long grassy shadow by the overgrown hedge. I feel triumphant. Another full moon rise. Each one different, each one extraordinary. Each one inducing a mild case of ecstasy. Actually not that mild. The sky stayed very light tonight, so why did the moon appear to become brighter in it? Bats.

• A full moon rises when the sun sets, and sets when the sun rises, because a full moon is on the opposite side of Earth from the sun. It appears to be doing the opposite of the sun.

• A new moon rises and sets at the same time as the sun because it is directly in front of the sun (so we can't see it).

6 JUNE | WANING GIBBOUS

Made a very long drawing trying to include the sun setting in the NW and the moon rising in the SE. The sky and the shadows in the field have to change as you move along the bit of paper. A flat piece of paper isn't really good enough. What I need is a circular room and a whole roll of paper. The inside of a grain silo or something, that I can attach my paper to. It feels marvellous to be standing here, imagining myself as the fulcrum between the sun and the moon.

16 JUNE | WANING CRESCENT

It rises in the small hours and sets at 16:30. I'm not really noticing it at all. The waning phase of the moon seems to pass me by. I've settled into a pattern of paying close attention to the waxing moon for two weeks and then losing sight of it for the next two weeks. The waning moon rises after sunset, and later each night, so it's only visible in the small hours. It IS up during the day, but it's more difficult to keep track of then because I'm busy doing daytime stuff.

- A waning moon rises before sunrise and sets before sunset. It is travelling through the sky in front of the sun, so its illuminated side is behind it - towards the sun.
- A waxing moon rises after sunrise and sets after sunset. It is travelling through the sky behind the sun, so its illuminated side is its front - towards the sun.

19 JUNE | WANING CRESCENT
The moon seems to have got stuck in the small hours. When I first started doing this, I confidently told Fred that the moon rose an hour later every night/day. I don't think I was right. It seems to be rising at the same time every night at the moment. Confusing.

23 JUNE | WAXING CRESCENT
21:40 Today, when I get the first glimpse of the scalpel-sharp new moon above the wood, it feels like I'm greeting an old friend. The whole moon is visible, but just the leading edge shines. The moon is simultaneously reflecting light directly from the sun AND reflected sunlight from Earth. It's called Earthshine – it's the solar system equivalent of looking in the mirror; it's proof that we exist. In the cosmic pingpong of light, the crescent moon is a pie chart in the sky, reflecting second- and third-hand light for the fourth time, right back at me.

above Earthshine

26 JUNE | WAXING CRESCENT
20:30 Perfect crescent moon in yellow sky. Mosquitoes. Bats. I try and find the moon's reflection in the stream and wonder if the enormous eel Billy saw today is also looking at the moon.

5 JULY | **FULL MOON**
Warm and clear. I think I'm becoming expert now. I know where the moon will rise and I've found the best place to sit to draw it. At the top of the big field, with the sunset behind me, I settle myself on my upturned bucket and set my pastels out. Moz is hunting for rabbits. I start drawing, noticing the last of the light on the tops of the trees, their yellow green fading to smoke. The sky remains light, and the trees remain light too.

Impatient, I check my watch. 22:20. I stand up. I sit down. I stand up again and I gasp. I get a glimpse of bright light behind the trees. As I watch, it reveals itself to be a pinprick of intense light. A pinprick? That's not the moon. I don't understand. Then the sky around it begins to turn dark orange. Idiot – the bright light is Jupiter.

above 5 July 2020

It appears above the trees, heralding the moon which I can now see clearly, broken up, behind the trees. The sky is now red, the moon is deep luminous red. I gasp (again), I almost clap, I sigh, maybe I whoop. Every time I look up, the light has changed. More yellow, there is a red band of rays stretching far above, and now there's pink. Gold, rose, russet, dark fire. And Jupiter. Bloody hell – I've left my orange pastel behind. I fumble around and find an orange pencil. I think it's orange. It's really dark now and I can't see what colour is what, but I keep drawing, guessing.

I have the best seat in the house and the universe has put on a show just for me. Or that's what it feels like. An hour of fever-pitch drawing in the dark. Of time racing. Of heavenly colours and light. And then it's all over. Things settle, colour steadies itself, the moon leaves the atmosphere behind, and I'm exhausted. Smell of crushed mint as I walk home.

16 JULY | WANING CRESCENT

I've bought a planisphere and a red torch. Apparently red torches are good for seeing in the dark without blinding yourself. A planisphere is a simple low-tech – 2 pieces of cardboard attached in the middle – way to plot the sky hour by hour for every day of the year. I want to be able to see the movement of the stars through the year and learn their names. I'm resisting downloading an app for two reasons: firstly, using a planisphere engages my brain more, I have to set it for the right hour of the right day of the right month; and secondly I have to work out the position of other stars to be able to pinpoint the one I'm interested in. It roots me here, now. With a little bit of brain effort I'm hoping the information is more likely to sink in than if I just point my phone at the sky and obtain an instant answer. It seems to me the app method will just perpetuate more of the app method. I'm hoping that the planisphere will actually teach me, so that eventually I will be able to look at the stars without using it. Using it less anyway. The second reason not to use an app is the glaring white light from a phone. I've learnt that a second of light from a phone can undo thirty minutes of dark adjustment in my eyes.

01:20 Wandering about looking at stars. I can recognise the Plough and the North Star. And Cassiopeia, I think. A vivid long-lasting green streak from S to N. The biggest brightest shooting star I've ever seen.

> Polaris, also called the Pole Star or the North Star, is the only star in the Northern Hemisphere sky that never moves. All the other stars and constellations appear to rotate round it in the course of a year. In addition, it's always at the same altitude above the horizon as the observer's latitude. The further N you are, the higher in the sky Polaris will appear.

17 JULY | WANING CRESCENT

Midnight, no moon, it rises at about 02:00. Looking at stars. The Milky Way is obvious. Jupiter in the S. Everything else is a confusing splatter. I've always resisted learning the names of the stars before. I don't really know why. Perhaps because it just looks so complex. A mountain of a task. I don't get much further than the Plough and Cassiopeia again tonight. I try drawing with the red torch. Like with a normal torch it's impossible to see colour. But useful for seeing where I've put things, I guess. 4 shooting stars.

19 JULY | DAY BEFORE NEW MOON

Warm, clear night. Andrew gave me a telescope for my birthday and we wait impatiently for darkness. In the field we inexpertly fumble around with lenses and lens caps and eventually point it at the brightest thing in the sky. Good grief – our hair stands on end. It's as if the eyepiece is burning us. We keep looking through it, and pulling away and making exclamations, and then looking again. We can't believe what we are looking at. We can see Jupiter…AND ITS MOONS! In the same plane there are two bright dots on each side of the planet. They must be moons? No!? The moons of Jupiter! From here!? Us!? From one minute to the next we have gone from tired couple, bickering in the dark over an instruction manual, to electrified space travellers. Our world has just expanded by millions of miles and all we did was stand in this familiar place and look through a cheap telescope. Buoyed by our success we aim our spaceship of discovery at Saturn, we speculate about the exact number of rings we're going to see, but are deflated when all we see is a fractionally less fuzzy blob than we can see with our eyes. Undaunted we look at stars, but everything is a letdown after Jupiter and its moons: all we see are sharper dots of light. There are lots of moving lights amongst the stars; I think they must be satellites. They are rather like mites in the hen house, which only come into focus when your eyes adjust to the gloom.

When I look down, each one of the tiny white flowers that make up the yarrow's inflorescences is starkly visible. They create a cosmos of petals round my legs. I feel quite dizzy with it all –

above 1 August 2020

Jupiter's moons within sight above me, yarrow petals ad infinitum around my shins. The universe is expanding, in every direction, every time I look.

> Planets do not produce their own light; they are illuminated by sunlight. They do not twinkle.

30 JULY | WAXING GIBBOUS
23:30 Wandering about in the big field with Moz. Can't bear to go in. Can't get enough of looking at this beloved, scrubby field in the dark. I'm simultaneously in a familiar and an unfamiliar place. A stranger in the place I know best. Rabbits thump the ground behind the brambles.
I go down to the gravel pits. The lustre of moonlit water is tarnished by pondweed. Complete silence except for muted, almost inaudible kissing noises from the water. What is it?
I lie down looking at the moon behind the poplars, listening to the kissing – wondering what's making the noise. Is it animal? Vegetable? Or mineral?

1 AUG | WAXING GIBBOUS
21:00 I'm sitting at the top of the big field. It's warm, scattering of cloud, birdsong petering out, light diminishing. God, it's beautiful. The different reds of the drying docks are becoming luminous against the darkening grass. The moon is large and low, in and out of clouds near

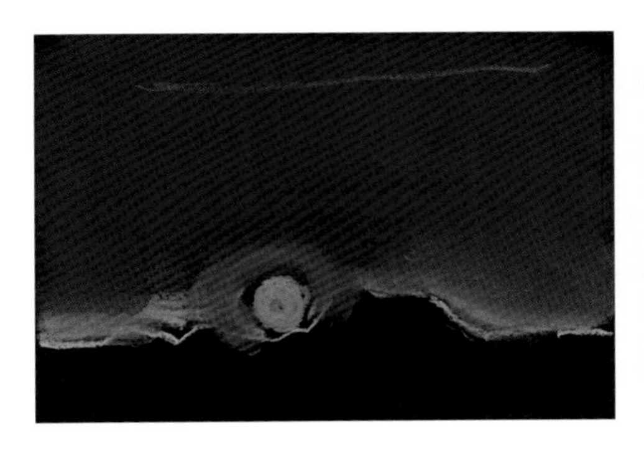

above & above right 3 August 2020

the horizon. Each unique cloud shape takes on moonlight in a different way. Moonlight shows up the depth in cloud. Pinks, so many pinks. Owl.

3 AUG | **FULL MOON**

Hot. Full moon rises at 21:09. I decide not to fumble around in the dark waiting for it to emerge from behind trees. I'm going to drive up to somewhere near Mile Road and get a really good view. I should be able to see SE well. When I arrive there's nowhere to stop, they're harvesting up here. Huge machinery with very bright lights everywhere. Drive around looking for good places to stop. No passing places have a good SE view. Drive around so long I see the moon has risen. Decide to go home. Stop in the lay-by halfway up the hill, the view is OK from here. I think I can open the boot and sit comfortably with my chalks next to me, but with the boot open the interior lights stay on. Annoying. Try and turn them off. More time goes by fumbling around with buttons that I've never pressed before and that don't seem to turn the lights off. What do they do? Eventually start drawing. This is quite comfortable. I won't lose my chalks in the grass. Then a car, with its headlights on full, comes down the road towards me, and it slows as it sees me. I'm completely and utterly blinded by its lights. The light is painful. Hellish. Terrifying. My eyes are burning. My head is throbbing. The car stops. I can't move because I can't see anything. I'm not sure which way anything is. The sudden immersion in thousands of unwanted lumens has stripped the world away. I'm paralysed and also praying that no one gets out of the car. I'm holding my breath, listening for a car door. Thank goodness it moves on. My eyes are still blind. And on fire. The space behind my eyeballs pounds and I'm realising how vulnerable I am. What the hell am I doing? It's 23:00, it's dark, I'm on my own sitting by the edge of a road, just waiting to be blinded and bonked on the head. I wait for the pain to subside and for some sight to come back. I begin to pack up but can't see anything because I can't turn the interior lights on. Oh well, at least all my stuff is somewhere in the car.

I get home with bruised eyes and feel an urge to bathe them in darkness, so I walk out into the field. I feel safe, darkness is acting as a balm. Slowly, patchily, my sight comes back. Clouds have blown in from the E, I watch the low moon behind them. Eventually I start to draw, it's past midnight, it's taken me three hours of pointless kerfuffle and I'm back where I started. Finally it's just the moon and me. And a moorhen screeching.

4 AUG | WANING GIBBOUS

Hot day. Last night's blinding has made me think about all the nocturnal animals that are run over. It's given me a greater sense of the speed, confusion and horror of their final moments. I'm finding it unbearable. It has also confirmed to me that drawing darkness with the aid of a light is not the way to try and do it. My eyes do work in the dark, they work differently from the daytime, but I can still see. I naturally adjust my speed and how I move around, but I can still do things in the dark. I realise that I'm comfortable in the dark. I'll just have to adapt my drawing. Not try and replicate daytime drawing. Not think of it as less, but as something entirely different.

We see in the dark through cells in our eyes called rods. They are more sensitive to blue wavelengths of light than they are to red. That's why blue flowers seem to come into their own in low light. Red light is used by airline pilots and in submarine control rooms, when night vision mustn't be compromised.

12 AUG | WANING CRESCENT

The Perseid meteor shower peaks tonight. 23:00 It's been a horribly hot day but it's cooled down a bit and I'm extremely happy to be sitting in a comfy chair, in the field outside my studio, in the dark, in expectation of great things to come. I have a big board, with a big bit of black paper taped to it, across my knees. I can name Cassiopeia, the Plough, Polaris. Is that Mars? Just above the E horizon? Silence except for crickets under the trees. I see a few faint silvery streaks. It's not till about 23:30 that the meteors become more visible and their trails have colour. I can't draw sitting in the chair, I have to put the board on the ground and crouch over it – looking up, looking down with an exaggerated sweep of the head. I'm amazed by the colours of the meteors, there are so many – glacial blue, yellow green, emerald green, scarlet, pink and silver. Why? I try and put them on my drawing, although I'm just guessing which colour I'm putting on. There are shooting stars everywhere, not just from the NE, and the show just gets better and better and more and more colourful. I'm beginning to feel really weird, though. My neck and head feel as if they might snap off, my head is throbbing. I'm longing to stop but I keep thinking – just one more – just one more – just one more fabulous one – an even bigger one... Eventually I give up and tiptoe to bed at about 02:00. My spine and neck unwind slowly, painfully, but it's a relief to be lying flat. I fall asleep knowing I'm a complete lightweight and I'm almost certainly missing the best bit of the show. I'm not a great fireworks lover – Bonfire Night is generally cold, it's chaotic and I don't like the noise. Andrew loves them, though, he wants the bangs louder and louder, he cheers for the ones that sound like bombs. The colour and sparkle never in my eyes make up for the cold, confusion or din. I realise that watching the Perseid meteor shower is perfect for me. I'm on my own, it's warm, it's beautiful, and it's in complete silence.

31 AUG
22:00 I'm in Cumbria and I'm confused with the direction of things. Mellbreak is blocking the entire S sky and the waxing gibbous moon is about to disappear behind it. I thought trees were annoying, but losing the moon behind a fell is worse.

1 SEPT
One day before full moon, in Cumbria. 20:30 Tracking the unbearably slow arrival of the moon behind Robinson. Although I can't see it, I can watch its progress W and up by the intensity of its glow on the sparse squiggle of clouds above the horizon. When I watch it emerging from behind trees at home, there's always a shocking chink of colour between branches to reassure me that I'm not making the whole thing up. Here, the fells are solid, there's no peeking. When the moon does climb into view I'm surprised by its size, though. It seems rather small. Weirdly my foliage-enclosed, mid-Norfolk moon viewing station seems to throw up much bigger moons. I guess this is my brain doing unconscious calculations to do with distance and tree/fell height.

If you measure a full moon, looking huge, on the horizon and then later that night, when it's high in the sky, you'll be amazed to find they are exactly the same size.

6 SEPT
In Cumbria. 20:30 Thin cloud. The waning gibbous moon hasn't risen yet. I walk up the track and then up the steep bit between the trees. It's really dark but amazingly I don't trip over once. The stones on the track are unexpectedly visible in the gloom. When I first came here, my father-in-law told me a story about Cumbrian rocks made – in part, he said – of starlight, and how the paths on the high fells glowed with starlight on moonless winter nights. I thought he was being fanciful, but he definitely had

a point. Although there are no stars or moon tonight, the rocks here do seem to glow in the dark. Especially in the particularly dark patch under the hollies. I always trip over the rocks there in daylight but not tonight. I didn't stumble once.

7 SEPT
In Cumbria, thick cloud. Warmish. 21:30 I walk up the track again, on the starlight stones. I continue up Mellbreak, to the top of the grassy path, and look back at the valley. I see below me the reverse of a starlit night. The land is the blackest of black and each house is visible by its lights. There's Thrushbank, High Cross, Crabtree, Godferhead… The spaced-out lights are like a constellation and the pub, with its concentration of many lights, is a nebula or something. A supernova maybe. Lorton is a background smattering of Milky Way. The sky is starless and pale, the lake is paler.

9 SEPT | WANING GIBBOUS
Home. 23:30 Cold, still. Deep, deep orange moon rising very far NE. It's on the N side of the big oak tree by the golf course track. That's about as far N as I've seen it rise. There's a very bright star above it and to the N. It's spitting and squirming and twinkling; red green, gold. And, what's that? A blue smudge? A star? A few stars? It looks like a tight group of stars out of the corner of my eye, but when I turn to look at it directly it becomes a smudge. Rifle shot. 01:45 My planisphere tells me that the star is called Capella. Wikipedia tells me it is actually 4 stars, not 1. It's 2 pairs of stars going round each other. Wow. I'm amazed by everything at the moment. Imagine if we had 2 suns? And the blob…it really is a group of stars, it's called the Pleiades. And they really are blue.

We view the sky through Earth's atmosphere. It's a relatively thin layer of dust, water and ice that lies round the surface of our planet. Because the Earth curves, it means you are looking through a lot more atmosphere when you look at the horizon than when you look straight up. All the stuff in the atmosphere does weird things to light. It blocks and bounces different wavelengths of light about in different ways. It scatters the blue and the green wavelengths back out to space. That's why the sky appears blue. And why the moon and sun tend to be red when they rise and set. It's also why stars appear very jittery when viewed through the atmosphere as they rise and set.

21 SEPT | WAXING CRESCENT
19:15 Clear. Perfect gold moon hanging in a peachy sky above the end of the lake. Makes my heart lurch; my old

friend is back. The more exquisite the sight, the more cack-handed my drawing. Crows, jackdaws, rooks, bats, traffic. Smell of plums.

28 SEPT | WAXING GIBBOUS
Broken cloud. 19:00 I have 4 small pieces of paper taped to a board, 2 black and 2 white. I have a tiny box of pastels – 2 pinks, 2 pales, 1 weird green, 1 dark blue grey and a black pencil. I balance the box on my board and I go into the field and draw the moon rising above the trees. It's pretty dark, I don't look at the paper much, just at what I'm drawing. I can guess which colours I'm picking up. I move nearer the fence and do another one. I do 4 drawings and go in to my studio. I laugh at the results but I love them.

above 29 September 2020

Mad, wobbly, lines and smudges, awkward, unclear and strange. These drawings are what it's like to move about and draw in the dark. They are honest and full of energy.

29 SEPT | WAXING GIBBOUS
20:30 Mist rising and falling over the field and round my studio. Cold. Smell of damp vegetation. In the SE the big, bright moon is wearing zebra stripes of dark cloud tonight. 21:30 The stripes have melted away. The sky is sharp and starry above the mist. So is the top of the birch tree. Misty droplets glitter all around me. Smell of crab apples. Owl.

1 OCT | **FULL MOON**
19:00 Full moon rising due E. Misty again. Cold. I'm roaming round the field with a board and several bits of paper stuck to it. This feels so much more natural than formally setting up in one position to do a drawing. I move about, finding the dark gold moon between trees, over the stream and, best of all, behind trees. Moonlight in the mist becomes 3D. A sort of solid spilling round trees and over the grass. There's no differentiation between the air and the ground. The birch tree and its shadow are one complete thing (a weird, black, triple-length birch tree banana); the moonlit air and grass is everything else. As I walk towards the field maple hedge, the moon cocooned in a half halo of gold, green and red suddenly bursts through. Two beams of misty light pour into the field. If I move to the left it disappears. If I move in any direction it disappears. It's just from this one spot that I can see it. I start drawing, but a couple of minutes later it has gone. Moving, everything's always moving. My drawings are funny – very insistent marks where the light is spilling through. Actually looks surprising like the startling burst of light.

3 OCT | WANING GIBBOUS
20:45 Waning gibbous rising E. Blood red, just behind Mars, which is red as well. Low mist. The moon has been rising in the same place and at pretty much the same time since just before the full moon. Why? It doesn't normally do that. It's having a strange effect on time – there seems to be much more of it. Time appears to be almost at a standstill. I can see the moon just above the curved hedge if I stand with my back to the kitchen window, and I can use the light spilling out to draw by. It's quite good, as I can see the colours I'm using. The light is very soft. Only trouble is my shadow, which obscures the board.

above 23 October 2020

16 OCT | NEW MOON
00:00 Clear, starry. I think I can see Betelgeuse. Very excited. I read all about it in July when it wasn't visible.

23 OCT | FIRST QUARTER
20:00 The moon is very low. Due S. Why is it so low? When the moon or sun are in the S, they are at their highest. Tonight it's barely skimming the trees. It rose at 15:00 and will set at 23:00. Only 8 hours above the horizon. Why? Blink and you miss it. It's the

same colour and shape as the ragged, yellow and orange dahlias caught in the light from my studio window. The flowers, so close to me, are almost the same size as the moon beyond them. I draw them again and again. My least favourite dahlia and the strangely low October half moon seem to have a thing going on. The moon has moved a long way W by the time I finish.

26 OCT | WAXING GIBBOUS

16:45 Pale moon rising in the SE. It's hardly noticeable next to an expanse of cloud to the S that the sinking sun has painted gold, orange pink and purple. 17:30 The sun and its brazen palette of colours has gone. Now the moon can thread its subtle colour through the twilights. The stream glistens liquid gold as I walk under the alders, and I hear an otter squeak.

31 OCT | **FULL MOON**

This is the second full moon in October. The second full moon in a month is called a Blue Moon apparently. 17:00 Cold and windy. Fat clouds bring heavy rain on and off. I dash out during a lull in the wet proceedings, run to the road and stand by the bins to find that the rain can't dampen this searing cadmium-orange blue moon. The wind blows enormous drops of freezing water off the trees onto my head. A large tractor with an empty trailer careers over the bridge and through the big puddle; its many lights obliterate the world. I look down, I try and shield my eyes. It rattles round the corner and I look up to see the moon is being devoured by a large, dark cloud. Only half visible. Only the N edge remaining. Gone. The guilty cloud has grown pink frills round its edges. I hear another tractor coming down the hill and a particularly big drop of water lands on the back of my neck and runs down my back.

You'd think that 12 months equalled 12 full moons but it's more complicated than that. It takes the moon 29.5 days to complete its circle round Earth. That's 354 days. 11 short of 365. Roughly every two and a half years, a 13th full moon will be included in a year.

1 NOV | WANING GIBBOUS

One day after full moon. 21:45 Incredibly bright. Clear and still. The moon still looks whole tonight, with a distinct clear-cut outline all the way round but for an almost imperceptible fuzziness on the top of the leading edge. 22:30 I'm standing amongst the giant black stems of the near poplars looking towards Luther's 3 Lombardy poplars, whose trunks are white in full moonlight. Muntjac barking. An owl. Smell of fermenting apples. There's dripping from my studio drainpipe. I can see the colour of Moz's coat in this silvery black world.

13 NOV | WANING CRESCENT
The moon has just set, it's travelling through the sky just in front of the sun. Although it's been a clear day I haven't been able to see it. 16:30 Civil twilight, I'm in the big field, the sky is deep blue, and Mars is very obviously red and bright in the E. Moz is looking very red, too. I'm watching the fading light on the autumn colours. The W end of the blackthorn clump is a pale lemony yellow that is refusing to dim. Crows. V's of noisy, low-flying Canada geese flying W, silent mallards in smaller groups flying E. When I can't see to draw, an owl starts up from Luther's yard. 18:30 I look at my drawings back in the studio. Funny, I seem to have missed one of the bits of paper. How did I not notice? Half the drawing is on the board.

17 NOV | WAXING CRESCENT
22:30 Clear, cold. Down here: owl, moorhen, muntjac, me and Moz. Up there: Orion, the Pleiades, Taurus, Mars. I'm getting better at finding my way round the stars.

28 NOV
I'm in Cumbria. 16:00 Without warning the pale pink moon is sitting on top of Robinson. 17:00 I'm sitting by the tarn; the moon and its reflection are to the left of Mellbreak and its deep red reflection. The water is very still and the reflected world looks like a shinier, darker version of what's above it until you notice the moon has been elongated into a sort of orangey pink cocktail sausage. After about half an hour of drawing I hear faint wing beats and the tarn explodes with light. The now bottomless black of Mellbreak's reflection is pierced by deep red darts. Shards of red briefly expand into circles and vanish. The contained cocktail sausage of moon is now a million shaking flakes of gold quivering from bank to bank. The ducks have detonated the underwater moon and the underwater Mellbreak is dissolving in light. As the ducks settle, the water and its reflections establish a gentle wobble until another unseen and almost inaudible bomb squad arrive and blow the whole place up again.

I finish drawing and just sit and watch the squadrons arrive, each red dart of light in the inky water a burst of intense pleasure in me. Moz has moved very close to me, I can feel her shivering. I'm getting cold, too. I walk/crawl away (a bit like I imagine they do in the SAS), so that I don't frighten the ducks.

30 NOV | **FULL MOON**
At home. 16:15 I'm standing in the middle of the big field. The pheasants have finished their garrulous nighttime routine, crows are still at it. Groups of ducks begin to fly over me. A heron shrieks. I turn round and the bottom of the field has disappeared in mist. Mist begins to rise in front of me. Mist undulates. There's Mars. I can see the

above 30 November 2020

moon's glow silhouetting the trees. I think I'm seeing the sky through two types of mist. There's the ground mist that is swirling around me and is plainly visible up to a certain height. And the sky above is misty, too. And now – there's the moon, it's huge and orangey gold, just above the trees. It has a double halo of red and a green around it. As it moves quickly up into the smeary sky, something catches my eye. Moonlight is slowly filling every puddle and water-filled hoof print in this boggy ground. The higher the moon rises, the thicker the mist appears to become. Alchemy of water, air and light. I'm very cold.

8 DEC | WANING GIBBOUS
Waning gibbous moon set in NW a couple of hours ago. 15:00 I'm at the top of the big field, by the young oak tree that the buffalo have ring barked. The mellow December reds and browns of spent seed heads, dead stalks and bare branches are succour for short days. I'm drawing the sun set behind the poplars. It's as fast-paced as a moonrise. Colour is not stable. Pin-rush steps forward, sparkles, glows, grows shadows and retreats. Not a breath of wind but this place is on the move. Or rather rearranging itself. Wrens in brambles behind me.

15 DEC | WAXING CRESCENT
Am realising how entwined the sun and the moon are. To understand the moon you have to understand the sun. In fact, if I try and talk about the moon to anyone (or even think about it to myself), I find I always have to start with the sun.

- Sun-day
- Moon day
- With the absence of a clock on the mantelpiece or a phone in your pocket, the most obvious marker of time is the day/night cycle.
- Bigger chunks of time can be measured (and plans made) by watching the moon and its phases.
- A fortnight is from the Anglo-Saxon fourteen nights (of course you would measure time by nights if you were watching the moon).
- A month (moonth) is the measurement of one moon cycle.
- A year is one Earth cycle.
- What we call time is the spin of the universe.

20 DEC | FIRST DAY AFTER FIRST QUARTER
I decide to plot the moon every night at 17:00 from now until full moon. I'll stand outside my studio, looking S, and draw the moon's position in the sky in relation to the trees on the horizon, to Mars, and to the stars, etc. Capella and the Pleiades are very visible in the E.

above Sunset, 8 December 2020

21 DEC | WAXING GIBBOUS
Winter solstice. Rained all day. Got soaked to the skin standing in a queue for half an hour outside the post office. All ten of us in the queue gave evil looks, behind our masks and under our hoods, to the bloke who had taken so long.

The great conjunction between Jupiter and Saturn wasn't so great from here; it was obscured by thick cloud. The moon isn't even vaguely visible, but the evening is lighter because of it. Which means I didn't slide in every single buffalo pat in the field just now. Plotting the moon's position in pouring rain is difficult.

24 DEC | WAXING GIBBOUS
No visible anything. Rain, rain, rain. Everyone is worried the river will flood the house. Frank was awake all night.

25 DEC | WAXING GIBBOUS
The river is almost as high as we've known it but luckily not in the house.

Everyone tetchy as no one slept well last night. The moon is high.

27 DEC | WAXING GIBBOUS
Sunny day. 15:40 Drawing the slippery light of the setting sun as it slides up the trees. The big moon is beginning to brighten behind the tips of the poplars which are still illuminated by the sun. It has a dark reflection in the flood water. Drawing the moon is intimate. It often feels as if the moon is only for me. I wonder why? I think about our human love affair with it, the romance of the moon, the way we can feel connected to loved ones who aren't physically with us by looking at it. Why don't we feel like this about the sun? After all, the same sun shines down on all of us. I suppose it must be because of the privacy, the silence and the romance of darkness. The fact we can actually look at the moon must have something to do with it, too. Our lives and rhythms are ruled by the sun, ours and everyone's activity starts with the sun and ends (well, for most of the year) with the sun. The moon is for free time, time outside the ordinary. The moon has to be noticed. I'm trying to ignore a strong smell of sewage which is slightly killing the romance.

Later, it's freezing, I'm in the field searching for the Pleiades in the brightness that surrounds the moon. I think I can just make them out. I turn to walk on and look down (I'm always on the lookout for buffalo pats) and there at my feet in the frosty grass is a collection of about ten golf balls. Thrown there by my boys, who use them as missiles for a sort of target practice game, they have been transformed by moonlight and ice crystals into a terrestrial version of the Seven Heavenly Sisters.

above 27 December 2020

29 DEC | **FULL MOON**
It was a really stupid idea to plot the moon at 17:00 every day, on one piece of paper – in December. Even on clear nights it's difficult to plot, by eye, with any accuracy. Once you leave the ground, and known objects, you're only guessing distances. In the rain you might as well not bother.

30 DEC | **FULL MOON**
Full moon rises 15:34. It rose behind thin cloud which got thinner and thinner until none was visible. The colour of the ring of light around the moon changed every time I looked. 22:00 Clouds are drifting in from the NE. When a particularly small cotton-wool puff passes directly in front of the moon it is transformed into a puff of rainbow and reflected, for a second, in the large puddle. I wish everything would slow down. Torture to see interesting things but never for long enough to draw.

00:00 The ring of light round the moon continues its never-ending changing – yellow, orange, dark red, green, and even turquoise. Why? Ice, water or dust particles in the upper atmosphere that I can't see? Perhaps the air is always moving in the mesosphere and the troposphere, presenting light with an inexhaustible number and configuration of particles to bounce off in different ways, causing me to perceive an unending change in colour? Whatever the reason, I'm beginning to understand that the more I look at darkness the more colour I see.

The native people of North America call the December full moon the Cold Moon. Well, that's true here tonight. So cold that there is no contest between loitering outside drawing and being inside my warm house.

A raven arrived here two or three days ago. Dwarfing the crows, its throaty cronk stirs in us a longing for the north. I hope my neighbours don't shoot it.

31 DEC | WANING GIBBOUS
07:50 Thick mauve frost in the field at sunrise this morning. I realised that the sun and the moon have swapped places, exactly, from their positions in June. Today the moon was setting behind the farm – in the NW, where the midsummer sun sets. And the pink sun was rising behind John's covet – in the SE, where the midsummer moon rises.

Recognising these moments of balance is satisfying. It's taken me a year, and I still don't really understand the motion of the moon, but what I've learnt is that the sun and the Earth are everything to the moon. Viewpoint and sunlight are all. When I can discern that all three – sun, Earth and moon – are aligned in some sort of midwinter synchronicity, well then, maybe I've just scraped a level one (definitely without distinction) in moon watching.

7 JAN | WANING CRESCENT
08:45 Frosty. Cold. Woodpecker drumming in the wood. The moon is above the poplars. It still has the faintest pink glow. The water buffalo stand about breathing out huge clouds of purple mist. We can all hear Edward's low loader coming this way with a new bale of hay for them.

They go mad; they jump and run in tight circles and they headbutt the bale when he lowers it over the fence. They do victory laps with mad headdresses of hay attached to their horns. Within 10 minutes the bale is flattened and they are covered in hay, settling down on it for a morning of munching in the sun. The now-bleached moon sinks behind the wood.

8 JAN | WANING CRESCENT
09:30 A pristine, frosty morning. Blue and white and black. Long cold shadows that won't budge all day.

22:30 Icy, foggy – everything is grey. I don't bother to get my drawing stuff from the studio when I go out. I feel like I'm moving around in a grey dome. The dark trees round the edge of the field mark the circumference and the divide between the sky and ground – there's no visible difference between the two. When I tip my head back and look up at the zenith, I can see a star – in a little window of inky black sky. It's like a stud at the top of the universe. It seems as if everything relates to it and the world is held in place by it. The light from the kitchen window makes little headway into the gloom, it's being pressed down. I decide to get my sketch book and try and draw the 360-degree grey dome with the piercing star at the top, but I can't make it work on a rectangular piece of paper. I can vaguely see what I'm doing – dark marks on white paper – but I can't see what I'm drawing. What AM I drawing?

Orientation, direction and the idea of the sky as a dome have become more important to me over my months of moon watching. My view, my way of looking at and of being in this familiar place, has changed immeasurably. I can see that there are precise places on the horizon and in the sky where the moon and the sun and the stars will be at certain points – either of the day, or the month, or the year. But (there always seems to be a but) those points change depending on where I'm standing. If I'm in the field by the house, the wood is W. If I'm in the big field, the wood is N and E. Every time I move, the sky dome and its compass-point directions move with me. Tonight, though, wherever I am, all the identifiable features arranged around the horizon are invisible. N, E, S, W are all the same indistinct grey. The only thing I know for sure is that that star points up.

The zenith is the bit of sky directly above you. Stars move through it.

10 JAN | WANING CRESCENT
Orion is magnificent tonight. Betelgeuse, Rigel, Bellatrix, Nair al Saif… I list the beguiling names as my eyes plot the pattern. So seduced am I by the colours and crystal glitter that my feet leave the ground and I think I'm actually soaring through the cold, blue-black infinity. Eventually,

though, my neck hurts and I have to look away. Space and time and the constellation fold back in on themselves. Unbelievably, minutes have passed – not light years. Here I am, standing by the veg patch, and there, in front of me, two very close-set eyes glint green and unblinking back at me from the darkness. Slinky has been standing there, staring up at me, all the time. The patience of my beloved, non-star-gazing hound gives me a sudden jolt of homesickness. His ridiculously close-set eyes are everything the cosmos isn't. They are warmth and safety – they are love. They are here with me now.

As I fall asleep later, I think about the distance between Sirius and Procyon, the Dog Stars, and the distance between my dog's eyes. And I also think about travelling between the two in a matter of seconds.

13 JAN | NEW MOON

I've got the hang of some very basic things about the moon. I know that it rises later every day. This rising later brings it round in a full circle, time-wise, every month. How much later it appears on subsequent nights/days isn't a set amount of time: it can rise somewhere between 1.5 hours and 15 minutes later every day.

I've also understood that the moon rises at a different point on the horizon every day – anything between a 10 degree and a 1 degree difference. Within a month it will rise and set in different places along the horizon, following the course the sun takes in a year. For instance, in January the new moon will rise with the sun – in the SE. It will then rise further to the N every day until full moon, when it will be rising in the NE, directly opposite the setting sun. In the next 2 weeks it swings back SW, a few degrees at a time, to catch up with the position of the sun for the next new moon.

It took me months to understand this much, and I still don't understand it properly. Why is it sometimes 1.5 hours later than the night before? And then the next week only 20 minutes? Why is it sometimes 1.5 hours later to rise but only 15 minutes later to set? The moon's movements seem unpredictable, and it's exasperating not really knowing why. There doesn't seem to be a pattern. Well, nothing that I can discern. I can find reams of data with moon rise and set times. They tell me exactly where on the horizon it will rise and set, too, but they don't tell me why. Without the data I can only have a good guess at where and when it will appear – and that usually ends in frustration, hanging around for ages looking at the wrong place on the horizon. Here, in this tree-lined valley, precise position really matters: 1 or 2 degrees out and the moon might be blocked by a hazel bush.

I've been observing and drawing the moon for a year and the more I scrutinise it, the more puzzled I become. I'm longing to make sense of it. What I need is a moon expert – but where do I find one? I've read books, which are either way too science-y and go off into outer space, or they're all about the Apollo landings. I've written emails to likely people and organisations, but they've all ignored me. Maybe my emails were too jumbled and lowbrow? Maybe I'm just plain stupid. I feel thwarted not knowing how to find information, which I know must be readily available. Perhaps what I want to know is so basic nobody ever explains it, because it's assumed everybody knows. I'm sure there's some wonderful person out there, keen on naked-eye moon observations, who'd love to tell me, in simple language, everything I want to know. If only I could find them.

Incarcerated by snow and in the absence of any help, I decide to try a different way. I look at the information I do have. I have columns and columns of numbers, for dates, times, direction, illumination and altitude. I'll try and visualise them. I'll make a chart.

I draw a circle in the middle of a big piece of paper for the points of the compass. I try and break it down into 360 equal degrees. More difficult than it sounds. So that's my 360-degree horizon line. I draw six increasingly large circles around my inner horizon line circle. Each circle is 10 degrees of altitude – from 10 to 60. Taking a random lunar month – from October new moon to November new moon – I plot the exact direction of the rise of the moon: I find the highest altitude the moon gets to that day and I draw it, in its right phase, in my 'sky'. I also plot every moonset in the same way. It doesn't really make sense because I can't incorporate time – so there's no idea of day or night. Unexpectedly a butterfly shape of moons emerges onto my piece of paper. I'm wildly excited – mainly because it's pretty. Does it tell me anything?

14 JAN | WAXING CRESCENT

More snow. More charts. Of course I have to draw more charts, I want to know how the months vary. I'm going mad drawing charts. My studio is crammed with them. I've changed my compass so that S is at the top. The compass is my horizon line after all, and no one would view the movement of the moon by looking N. Very excited to draw charts for the summer and winter solstices and the September and March equinoxes. What exactly are the differences? Well, after all that frenzied work (each chart takes ages) I can now see that the solstices are exactly the same but the other way round…derrrr. The winter solstice month has the full moon at maximum height for the maximum amount of hours, and vice versa for the summer solstice month. Irrespective of the time of year, in each month the same butterfly shape appears. I think this must be the path

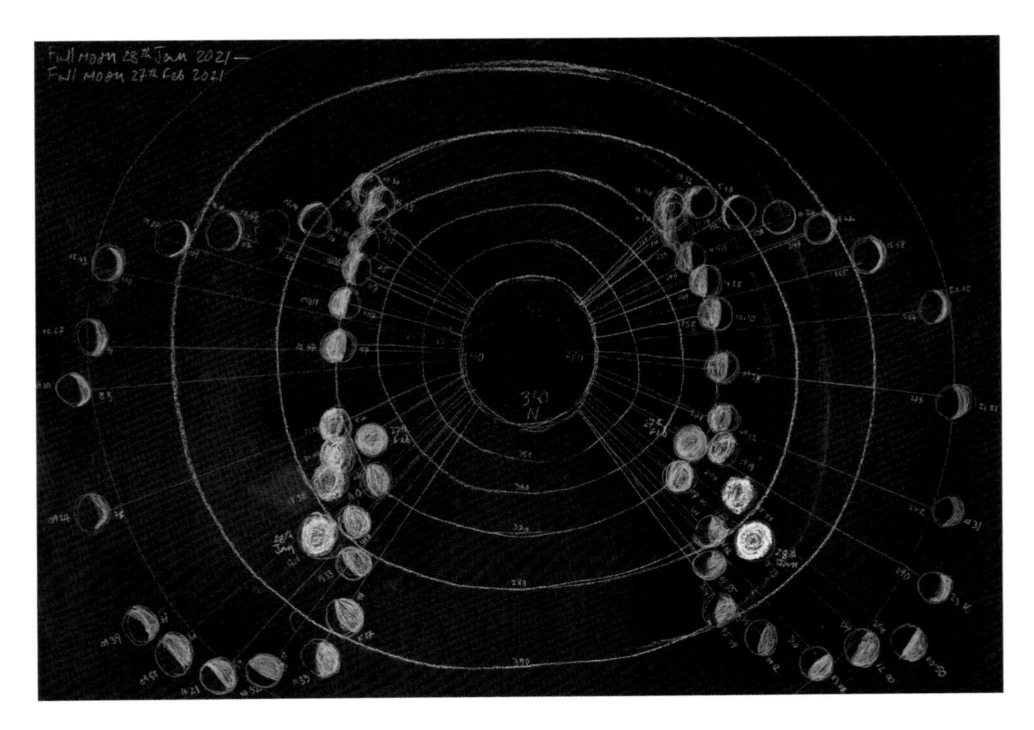

above January–February 2021 full moon cycle

of the orbit of the moon round Earth? Or is it something to do with Earth's spin? Or both?

Which direction to put at the top of my piece of paper is an interesting question, now I come to think about it. Why do I think N is the natural top? The Mappa Mundi, made in the 1300s, has E at the top – so the N-at-the-top idea must be a fairly recent thing. Another example of the limitation of a flat piece of paper when trying to represent everything.

15 JAN | WAXING CRESCENT

More charts. I add sunrise and sunset to them. I'm desperately trying to add time of day as well. I can't make it work, though. It just gets incredibly muddled.

I dream about circles all night. The moon circles on my charts. Unfilled circles, half-filled circles, quarter-filled circles, filled circles. Eventually just before dawn the circles turn into sliced potatoes – which I layer into a dish with butter, and put in the oven. I then fall into a deep untroubled/uncircled sleep for about 5 minutes before the alarm goes off.

17 JAN | WAXING CRESCENT

19:00 Moon high and misty pink, no stars. Moonlight on tiger stripes (leopard spots?) of half-melted snow on the long grass of the field. As I walk across it, the snowy patches are hard and crunch underfoot; the dark grassy patches are soft and my foot sinks into them. Try and draw it. It veers into a purple spangled horror of a picture.

We never get snow for long enough to really get to grips with drawing it properly.

18 JAN | WAXING CRESCENT
06:00 It's still quite dark and the sky is clear, so I go outside with bare feet to see the stars while the kettle boils. The sky looks unfamiliar, just a random scatter of stars. It takes me a while to work out where everything is. I've really only been looking at the sky in the evening and night. The morning sky is topsy-turvy. The Plough is upside down and Orion has gone. Venus is very bright in the SE. Cold feet, cold wind. What I'm doing, watching the moon and stars from here, without a telescope or binoculars, is called 'naked-eye observation'. Unwisely, this morning it was naked feet as well.

21 JAN | WAXING GIBBOUS
It has rained relentlessly. Abysmal weather for moon watching/drawing, it's just a lighter smear in the grey smeary sky. The water piling over the sluice is a deep thundering backdrop to everything we do. You can almost feel the violence of it through the floor when you're in the house.

22 JAN | WAXING GIBBOUS
23:00 Broken cloud, the moon is starting its descent in the WSW. It's silvery tonight, with a blue glow around it. The glow turns greeny gold if I look at it for long. As the thick chunks of cloud pass between me and the moon, it's as if I'm under a slowly flickering lightbulb. Fat drops of water glitter on everything when the moon comes out from behind a cloud. Fairy lights on each barb of the fence. They burst coldly against my legs as I walk past spent yarrow stems and drench me when I walk under a particularly laden Scots pine branch. I don't know which star is which in such a broken sky. I can hear the teal on the gravel pits when I'm by the gate. The weir is still roaring.

23 JAN | WAXING GIBBOUS
22:00 Clear, still, very cold and starry. The moon was visible all afternoon and is very high and bright in the S now. Moonlight is hazy, it has fuzzy edges, but the moon, it's prissily precise. Its perfect round edge is sharp. That's the point. It's in space and it's spherical. My moon's edges are the opposite of exact. The harder I try, the less circular they look.

24 JAN | WAXING GIBBOUS
22:30 Bright moon high in the sky shining through clouds. It's not freezing but a sprinkling of snow on top of every buffalo pat is the only reminder of today's wintery morning. Each randomly placed pat is glowing in the moonlight. This is the first evening I haven't stepped in one.

25 JAN | WAXING GIBBOUS
15:30 Meant to draw the moon at dusk but drew the sun instead. Through the dark poplar stems the green ball is extending its green light along the tops of the trees on the far horizon. Everything vegetable is drab and dark, the sky is colourless. It looks as if the weak January sun has fallen to Earth and is giving it an electric shock.
Teal.

26 JAN | WAXING GIBBOUS
22:00 Clear, still, very cold, starry, fox barking, owl hooting. The smoke from the chimney is illuminated by the moon. It's going up, in a more-or-less straight line, between the moon and Betelgeuse. It simultaneously makes the moon look nearer and further away. Weird.

27 JAN | DAY BEFORE FULL MOON
Hazy clouds made it a beautiful day and moonrise is very lovely. Everything is pink and mist is rising. A blackbird and an owl duet – one singing out the day and the other singing in the night. I've got a board with several bits of paper taped to it. As I'm only planning to draw in the garden tonight, I've got all my chalks with me as well. The day before full moon really is the best (for a moon artist). It's low to the horizon, it's beginning to glow, and I can still see my paper and my chalks. Looking NE I draw a group of jackdaws sitting at the top of one of the poplars. They are above the moon and above the mist. They are silent and still. Like me, they are looking. When I turn round to look W, the mist is rising up like a huge phantom over the house. Dark fingers curling into the pale sky. I walk back towards the ghastly apparition to find that, from within, the moon is transformed into the sugariest ball of light I've ever seen. Battenberg cake yellow and pink, with a polar blue thrown in. I draw the moon balancing exactly on top of a pollarded ash limb. It looks like a torch and its reflection is snagged in the swollen stream below. It's quite dark now but a robin is still singing. Direction and light are everything. Observing the mist in front of a sky still glowing with departed sunlight is a dark and menacing sight – but, although cold, inside the mist you find it's a dreamy place to see the world from. The dark mist turns out to be nothing but a sugar coating.
Teal.

above 27 January 2021

YEAR 2
JANUARY 2021-22

YEAR 2

28 JAN | **FULL MOON**
Moon not visible. Sound of water deafening. Miserable night. God, I was lucky last January when I set out to draw the full moon rising. I now know it was a complete fluke that it rose exactly where I was looking. I thought I was looking E, but actually at the road bridge the stream turns slightly N so I was looking NE.

I'm sick of hearing about the Wolf Moon: the names the press come out with are Native American people's names for the moon. Wolf because it's the time of year, in certain parts of North America, when wolves are howling. It's a local name, connecting people to their place on Earth, to their place in the universe. It describes the moon there. Not here. The January full moon hasn't risen to the sound of wolves howling in the UK since at least 1680, when the last one was shot in Scotland. No wolves here. In fact, calling it the Wolf Moon just highlights our estrangement from the natural world.

I'm going to make a list of possible mid-Norfolk January full moon names. Suggestions: mist, water, fog, cloud, rain, teal, cold, low sun or sun-so-low-it's-always-in-your-eyes, pale, frost.

29 JAN | WANING GIBBOUS
06:45 It's a clear morning. Blackbirds are singing. Stars are beginning to disappear. Last night's full moon is looking magnificent in the NW, it's just about to set behind the wood. It's as orange as the ring round the blackbird's eye.

The shortage of colour words in the English language is becoming apparent. I've been searching for other orange, pink, red words to describe the moon. I've been looking out for things that are similar. But it's difficult because the moon is light, it's luminous. It's not flat colour. A carrot is orange, but it's nothing like the orange of a rising or setting moon. My list so far is; Golden Syrup, maple syrup, marmalade, various flavours of jelly, whiskey, brandy, rosé, amber, garnet, cornelian, topaz, old engine oil, red diesel, Cherryade, Lucozade, Tango, a pochard's eye, a tench's eye, and (this one's weird, I saw them in the butcher last week) some internal organs. A lot of things on this list are transparent: you see them with light coming through them. The eyes, they are alive. And the innards, well, they're probably even pinker when they're living. The colour of the moon and anything remotely like it is to do with reflection and transparency.

17:45 Moonrise behind broken cloud. 21:00 The moon is none of the above at the moment. It's very bright, silver with a defined edge against black sky, and the clouds in front of it are really, really pale. How the hell do I draw that? Feeling pathetic, I don't even try. Go in and watch TV.

30 JAN | WANING GIBBOUS

18:00 The fodder beet in the field on the other side of the crossroads was harvested yesterday. Tonight I can see two tractors up there. Weirdly, from where I'm standing, the array of bright lights on them look like two big half moons. A pair of moons heading S, towards Hingham, one behind the other. The wind is from the NW, so they are soundless to me.

1 FEB | WANING GIBBOUS

22:00 I'm not drawing anymore tonight. I'm just standing still, enfolded in darkness. The moon is rising in the E, over the bridge, behind a thin shimmer of cloud. It's freezing. The moon is a gorgeous latecomer to the starry party. Pinky orange and monumental. Tonight it feels like the

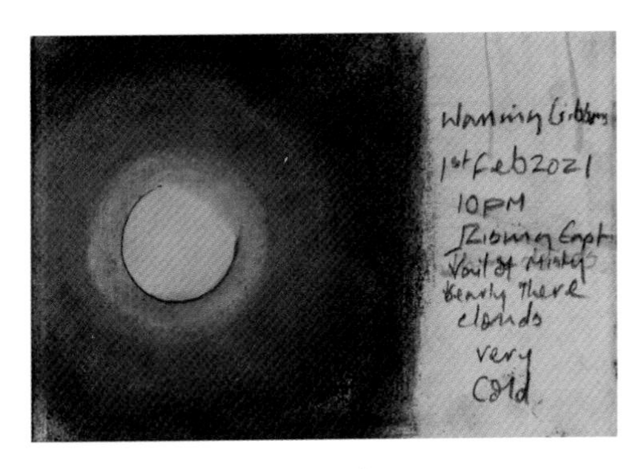

most beautiful moon I've ever seen. The missing leading edge is somehow poignant and perfect. The roar of water accompanies its ascent. To draw it would mean taking my eyes off it, and tonight that would be blasphemous. The moon is visible for such a huge amount of time at the moment. Around a winter full moon is the time to get your annual moon fix. If I stare at the big bright moon and then look anywhere else in the sky, a white imprint of the black tree-d horizon is splashed across it.

2 FEB | WANING GIBBOUS

07:45 The moon looks smaller this morning than it did last night. Partly because it's higher in the sky but mainly because the sun is just above the horizon and the moon is losing its glow. The missing leading edge seems bigger because there isn't a glow spilling around it.

23:30 I must have got the moonrise time wrong because I can't find it. It's depressing wandering about in the dark looking for the moon. I go down the road to the meadow, but it isn't visible from there either. Decide to go to bed. One last look from the garden and there it is rising directly out of the stream, dark orangey red – like a rudd's fins. Missing edge first. A bigger chunk missing, a sadder sight.

Don't start anything you want to grow during a waning moon. Don't start a journey. Don't plant seeds. Don't buy shares. Don't give birth. Don't get married. Don't start writing a book. Do, though, die and have funerals.

3 FEB | WANING GIBBOUS

08:00 The sun and the moon are both visible this morning. The orange sun is rising in the ESE into a bright bleached sky, the fading moon is setting in the WSW, in a much darker, bluer sky – as if it's still wrapped in night. The woodpecker is drumming in the wood.

4 FEB | THIRD QUARTER

It rose at midnight and set at 10:30. I didn't see it. 17:00 Civil twilight. Crows massing, red kites coming from every direction and joining a thermal swirl. Gunshot. Crow chaos.

5 FEB | WANING CRESCENT

07:00 Windy. Large sheets of cloud and the odd patch of clear sky. The pre-sunrise colours are blowing W with the clouds. They start in the place I expect to see the sun rise in about half an hour, and they are spreading W only. Not in a uniform arc. Weird.

21:00 Light night. Jackdaws making a noise around Luther's yard. Crows on every point of the compass. Friday night church bells faint but audible.

6 FEB | WANING CRESCENT

16:50–18:00 Clear, cold, still. Drawing the light fade by the gate into the big field. My back warm against a new silage bale. The noise of corvids, ducks and geese swells in volume relative to the fading of the light. The volume switch is being turned up, the light down. Light 5, volume 4. Sweet smell of silage. Groups of corvids going in every direction. Some perch on the top of the poplars, some in the wood. It's definitely a gathering. Separate groups are arriving. Many are congregating on the field on the other side of the lake. Light 3, volume 7. A steady flow over my head. The ducks on the lake are getting louder to compete. 3's and 4's of greylag geese take to the air honking loudly. Squadrons of teal in the air. A silent, sideways woodcock. Then a sudden whoosh as the raucous field gathering of massed corvids rises and descends into the ivy-clad poplars. Light 1, volume 10. Their noise drowns out the ducks. My spine is tingling. An hour-long music and light spectacular.

It seems to be crows and rooks in the wood, jackdaws in the poplars. And finally an owl. And it's dark.

8 FEB | WANING CRESCENT

07:00 I can see the barely-there crescent moon out of my bedroom window as I drink a cup of tea in bed this morning. It's in a gap in the branches behind the lake. I've never seen it from bed before.

10 FEB | WANING CRESCENT
18:00 Clear sky. The wind has dropped. Has the pressure dropped as well? Someone is burning plastic under cover of darkness. The smell is everywhere. It's so terrible I can't stay outside. My hair smells of it.

12 FEB | NEW MOON
23:50 It's freezing, -5 and the wind is in the E. Orion is low in the SW. It's too cold to linger outside. I always thought I was pretty intrepid at drawing outside in the cold. With years of practice I've perfected the art – layers of hats, coats and trousers, leather walking boots with two pairs of socks, which my trousers are tucked into. Waterproof trousers are good, too, especially for keeping the wind out. I thought I could tolerate the cold well enough to go out whatever the temperature. But at night the cold is another thing entirely. Sharper, piercingly sharp, it's excruciating. Norwegian duck down – no barrier. Two hats – meaningless. A convoy of huge agricultural machinery with flashing lights moves slowly along the main road.

14 FEB | WAXING CRESCENT
21:00 A sunny, cold, windy day has given way to a breathless, cold, clear night. Shadows are very sharp and black. Stars very bright. A pall of weed killer hangs in the air. I noticed 4 different sprayers out today. An owl squeaks and pheasants set each other off in every direction. I feel sorry for all the creatures that have no option but to stay out in this chemical fog.

15 FEB | WAXING CRESCENT
The moon is striking. Tonight is the first time I've seen it for a while. Time drags when the moon isn't visible.

Civil twilight (17:07–17:42): The moon is a pink sliver, with a pale ring of light round it, in an intense blue sky that appears light but in fact must be quite dark. Owl, crows, revving machine, 1 rifle shot.

Nautical twilight (17:43–18:22): The sliver of moon is now silver, with a blue ring round it, and the sky is undeniably dark now.

Astronomical twilight (18:23–19:02): The moon is golden pink now, with a pink ring around it that's darker on the outside. The sky is very, very dark. Owl.

18 FEB | WAXING CRESCENT
22:30 I can't quite see NASA's Perseverance rover landing on Mars tonight, but I can see Mars. It's very close to the moon. The Pleiades and Taurus are only just visible in the moonshine. There's Orion but even with a lot of squishing it in I couldn't fit Sirius into my picture.

Two military jets rumble around, blinking bright red. And in Luther's yard, on planet Earth, a tawny owl hoots.

20 FEB | WAXING GIBBOUS

20:50 Warm, still, clear. A very big red orange bright star rising in the E. What star is that? Very very big. 21:10 I go to the road to get a better look and it's gone. There is a reddish star rising in the NE but it's much lower in the sky and not nearly as big or bright. What the hell? I go back to the field. It's not there, in the gap between the poplars and the alders. If I move to the left it doesn't appear, nor if I move to the right. Then a whipping noise races up behind me through the tops of the trees in the wood, over the grass. It blows my hair forwards over my face and pushes my coat flat against my back, and whooshes on over the grass and into the spinny in front of me. Everything is still and silent as before. And that star is still missing.

22 FEB | WAXING GIBBOUS

Nice fat moon tonight. 21:30 I'm taking pictures with my new phone. I've been very excited about the fabulous camera on it. It takes amazing nighttime pictures apparently, but I'm appalled by the photographs I'm taking. It turns night into day, and an airbrushed day at that. I now know where some of the night paintings I've seen online recently come from – they are directly copied from new iPhone photos. I'm more than ever impressed with my own eyes/brain combo. My (and I mean every single sighted human's) ability to see the bright light of the moon in focus while also clearly seeing the world around me in subtle moonlight is something that Apple's iPhone optics can't replicate and definitely don't make better. I put the phone in my pocket and draw the moon reflected in a minuscule puddle that appears to be fathomless tonight.

Sidetracked by Wikipedia's list of measurements of various light levels

- Direct sunlight 32,000-100,000 lux
- Full daylight (not direct sunlight) 10,000-25,000 lux
- Overcast day, midday 1,000-2,000 lux
- Sunrise or sunset on a clear day 400 lux
- Dark limit of Civil twilight on a clear day 3.4 lux
- Full moon on a clear night 0.25 lux
- Half moon in a clear night sky 0.01 lux
- Moonless, overcast night sky 0.0001 lux

23 FEB | WAXING GIBBOUS

21:30 A cold grey night and an extraordinarily huge halo around the moon. The moon is quite high in the S and the halo seems to take up the whole southern sky, right up to the zenith. It's as if the moon is at the centre of a hole in the sky. A hole that it has made with its brilliance. The rest of the sky is grey.

Inside the hole, the sky is a beautifully rich green black. The golden moon has 4 long rays emanating from it. They are black gold and the ends are razor sharp against the edge of the halo. Inside the 4, equally spaced, rays the moon is wearing a crown of gold thorns. The outer edge of the halo is soft, fluffy and pale. Astonished. Dumbstruck. I've never seen anything like it. The longer I stare, the more insistent the pulsing feeling becomes. It started in my chest and is moving up to my head and down to my toes. The enormous thing in the sky, this phenomenon, is having a physical effect on me. 5 minutes ago I was on auto repeat, sorting washing, folding pants, pairing socks. I walked outside into a visitation from another world and now I'm wondering if I'm going to cry. Can I move? Should I move? Keep breathing, Tor. Will this thing evaporate? Is it growing? My grey, muddy, February field is a place of miracles. Eventually the tingling subsides enough to try and draw it, but it's off-the-scale huge for a bit of paper. Human hands can't draw heavenly wonders. It's even difficult to photograph in its entirety. Even with my new phone. The dogs just want to go back inside. A toad chirping by my feet.

00:30 The halo has faded. The moon is back to normal, the cloud has gone. I've found out that it was a 22-degree halo. They are not uncommon apparently and are formed when light is refracted through ice crystals in cirrostratus clouds. They often appear around the sun but can be difficult to see because of the glare. Oh. Is that all? What it forgot to say was that coming across one unexpectedly will blow your mind. These light phantasms affect humans in ways that are much more difficult to explain. The moon is the moon, a lump of rock tied to the Earth by gravity, reflecting sunlight, just going round. We, though, with our complicated brains, standing here, gazing up at it, through clouds and atmosphere, can give it any meaning we choose. Why 22 degrees?

The moon rose at 12:27 NE and sets 05:55 tomorrow morning NW. That's a whopping 18 hours in the sky.

25 FEB | WAXING GIBBOUS

Almost full moon. 16:50 I go to the far field to draw the sun set. After my January sunset drawings I've decided to try and draw a sunset for each month of the year. I wonder if the sun appears different at different times of the year? Do the colours of sunsets change throughout the year? Because of the way the Earth tilts, it means that here, in the Northern Hemisphere, we are actually further away from the sun in the summer than we are in the winter. It feels the opposite, though. Maybe that will affect it somehow? Perhaps this is a stupid thing to do. After all, everybody knows not to look at the sun.

This afternoon's sun is coming down to Earth in the gap where all the trees were cut down. The slash of light on the grass is acidic. The sun is quite a dark yellow and appears to be squashed into an oval shape as it nears the ground – as if it's resisting going over the edge.

22:30 Clear night. Not too cold. Very bright moon. Perfect drawing conditions. I sit on a chair on the studio step with low light shining through the door. I've only turned on the useless light above my desk. For the first time ever, I think it's perfect – not enough light to see anything properly but enough to see where my chalks are. I'm really enjoying doing little moon portraits. If I look at it through my glasses, it's silvery and sharply outlined – I can see the geographical features. Without glasses, it's golder with a blurry double red green outline. I prefer it without my glasses. It's so much lovelier and more draw-able. Why the added colour without glasses?

My heart sinks – I can hear the fut-fut-fut of the gamekeeper's little utility vehicle. He's coming this way. Looking for a fox to shoot on a moonlit night, I expect. Huge relief when no shot is fired and he eventually fut-futs away.

Geese are flying in the moonlight. Mallard, teal, moorhen call on the lake and ponds. Owl.

26 FEB | DAY BEFORE FULL MOON

16:30 Clear, cold wind. The moon rises an hour before sunset and gets steadily brighter as the sun sinks on the opposite horizon. I move from road bridge to garden as it climbs into the sky. Briefly the moon sends a golden tentacle under the bridge. The water under the whole length of the bridge is sparkling, there's light bouncing off the top and the curved sides. Moonlight is contained and multiplied, and the people in the cars driving over the bridge don't know the wonder beneath them.

18:30 I walk to the edge of Luther's yard and look back down the stream. The dead and pollarded ash trees are blackly silhouetted against the dark sky. Depending on where I stand, I can make the moon appear and disappear; I can snag it in their dead limbs.

27 FEB | **FULL MOON**

Full moon rising 17:38 ENE. I go to the meadow again and watch the moon rise to the N of the stream. If I was in the field next to the house, it would be rising somewhere behind the spinny

above 27 February 2021

and the neighbour's house; I'd have to wait another 20 or 30 minutes before I could see it. Composition-wise, this is a rubbish place to draw it from, but at least it's visible. January's full moon rise in the low was more satisfactory. February's full moon is soft metallic pink in a light, watery sky. It's heartbreaking. A perfect pink O. Within minutes, though, it's the colour of coral. There's a red haze around it. Overwhelmed by pure colour. How fortunate am I to be in a world where this is commonplace? As it pulls away from the horizon I walk back to the garden. It's golden now and visible behind trees; its reflection fits snugly into the space where a chunk of stream bank recently washed away.

21:15 An extraordinary moment standing on the edge of the stream. Moonlight has turned the thick overhang of ivy into flowering May blossom. I swear it has, I can smell May blossom. The ravishing reflection of the dancing, jittering golden moon jumps out of the whole length of the stream and enters the corner of my left eye. I hear birds singing.

What? I rouse myself and rush to my studio for a pencil and a piece of paper, but when I get back to the stream the moon has moved and the light has changed – I'm standing by a gloomy pool under a thick overhang of black ivy.

It sounds weird and totally stupid now I write it down. What I saw and what I knew I was looking at were not the same. It felt as if I had entered another reality. Momentarily it seemed obvious that the world has countless realities and that the moon can reveal them. I can't get the light hallucination out of my mind. I realise how long I've been out when I see Arcturus high in the E: it was wriggling and squirming through the atmosphere in the NE when I went out. Nothing stays still, everything is moving. I feel the spin of the Earth.

If you follow the curve of the handle of the Plough out into space, i.e. away from Polaris, you will come across very bright (slightly gold) Arcturus. 'Follow the ARC to ARCturus' is a good way of working out where it is. It's in the constellation Boötes.

Suggestions for names for the February full moon: robin, magic, hare.

28 FEB | WANING GIBBOUS

15:00 A cock robin in the hedge outside my studio turns towards the low sun and its breast is the colour of the moon the evening before.

16:45 Drawing the sunset, listening to the machinery on all points of the compass come to a stop one by one. A noisy squirrel didn't shut up, though. I know I shouldn't look at the sun but it's fascinating. It appears to eat trees and gate posts as it passes behind them. It sort of vaporises them, turns them red, turns them green, and leaves them black.

19:20 The moon is an immense blood orange. It's so much more difficult to draw it from total darkness. Moonlight slowly floods over the landscape, shadows strengthening and moving. Always moving.

Composition still bothers me. If I was drawing something in daylight I wouldn't put it in the centre of my drawing. Drawing the moon, when there's only black foreground, means I always seem to put it in the centre. God, it annoys me but I keep doing it.

2 MAR | WANING GIBBOUS

03:30 Very light. A mallard woke me up with an insistent quack, quack quack that went on and on and on. Like a car alarm. Got up and went to marvel at the tarnished gold moon, now in the SW and on its way down. The bloody mallard just went on and on. 17:00 The mallard has started up again after a daytime lull.

4 MAR | WANING GIBBOUS

07:30 Moon setting over the willow tree between Luther's yard and the wood. It's fading from the waning edge. The crisp edge facing towards the sun is still shimmery pink. The geological features are the same blue/violet as the sky. It looks holey.

5 MAR | WANING GIBBOUS

22:35 The moon hasn't risen yet. It rises at around midnight tonight. The starriest sky I've ever seen. Shooting star, low and silvery and slow in the NW. Owls. Pleiades glitter. Freezing. Beautiful.

This feels like the furthest my eyes have ever seen. From my wooded horizons, over and over again, I'm now seeing, clearly, right into the Milky Way.

6 MAR | THIRD QUARTER

I'm obsessed with drawing sunsets. It's dangerous territory. Sunset and chocolate-box go together like the full moon and lunacy.

10 MAR | WANING CRESCENT

I found a dead sparrowhawk today. Its eyes are dull but they are definitely orange. I imagine if I was to stare into the eyes of a living sparrowhawk they would be fiery. That's another one for my colour list of things that resemble the moon.

11 MAR | TWO DAYS BEFORE NEW MOON

No visible moon. Clear night. I went out after a long Zoom meeting and all day in front of a screen. I lay down and looked up, savouring the slow adjustments my eyes made from electric glare to heavenly reality. First nothing but the brightest stars, then as they got brighter they were joined by others, and then a whisper of smudge which soon became a real smudge, and then finally the Milky Way came into focus. I felt every muscle relax and vowed never to look at a screen again.

15 MAR | WAXING CRESCENT
18:00 I'm standing by the gate talking to the buffalo, watching the sliver of new moon becoming visible as the sun sinks further below the horizon. I realise the buffalo's horns are crescent moons. On the left side of their head they have a waxing crescent and on their right a waning crescent. The tawny owl is making warbling noises. The six moon goddesses burp the sweet smell of silage at me and drift away into the gloom. Duck and geese make a noise until late.

16 MAR | WAXING CRESCENT
21:00 The moon has set, it's very windy and there's some cloud. But I can easily see. At the same time last night, with a much clearer sky, it was inky black. The darkness was so black it was physical. Why?

20 MAR | WAXING CRESCENT
Spring equinox. Moon waxing crescent – day before first quarter. The moon is high, very high, and above the poplars. Edward said it was a cold, late spring so I guess it must be official now. It's certainly cold today. It was still and cloudy all day but it cleared by late afternoon to reveal a watery sunset just north of the stream. Orange, pink gold above a blue grey bank of cloud. The lake is thronging with frogs and

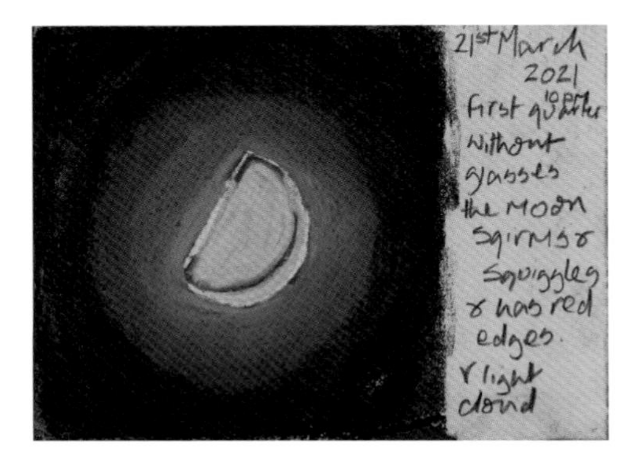

toads. The warty old toad has the most beautiful voice, like a far, far away duck – but sweeter.

I've lived here for 30 years and only just noticed that when the setting sun crosses either N or S of the stream it's a big moment. The stream runs perfectly W/E (with a slight N turn at the road bridge). The world has tilted towards summer when the sun sets to the N of the stream and when it begins to set to the S, winter is on the way. I couldn't have a better marker of direction. The Blackwater River is a sort of liquid sun dial.

21 MAR | FIRST QUARTER
22:30 Very light. Cold and still. Moon west of Orion. All that is seen of an aeroplane is a flashing light moving through Leo and Cancer. As it flies through Orion, under Betelgeuse, its contrail is suddenly visible and silvery in moonlight. Geese far away and briefly a blackbird. A car on the road with loud disco music. Silence, aware of the inside of my own ears, of listening.

Even on a clear night when the moon is high in the sky, the colours of it and of its halo, and even of the sky around it, constantly change. If I was to make a drawing of the moon every 5 minutes, from rise to set, each drawing would be different.

My night sky book tells me now is the perfect time to look for Praesepe in Cancer. It's an open cluster. Merely a smudge with the naked eye, but a group of dozens of stars with binoculars. I find it. My hands shake and it still looks slightly fuzzy. I have to admit that I don't really care about focusing in on any one spot – what makes me gasp is standing under the horizon-to-horizon dome of heaven. A few more dots don't enhance the wonder for me. The astonishing thing is that I'm here on the dark side of a planet, quite at home, feet in the grass, looking at the universe through my own eyes and my brain is fully engaged, trying to make sense of it.

23 MAR | WAXING GIBBOUS

14:00 Waxing gibbous moon rising in the E and reflected in the gravel pit. I lie on my front, on the edge of the water, watching the orgy of frogs and toads. It's their songs that have brought me here. I've been trying to find time to do this for days. It's a once-in-the-year event, the massed choir of hundreds of frogs and toads singing love songs. The croaky, purring frogs and the chirping toads in this old gravel pit make the sweetest music in the world. Amazingly I can draw the moon from here, but the drawing is hopeless; I'm too distracted by all the movement in the water.

24 MAR | WAXING GIBBOUS

19:15 It's a very bright, light night. It's warm, too. There's a bat. 23:00 I can see the dogs perfectly out of the corner of my eyes, but they disappear when I look straight at them. It's a weird sensation – I know it will happen but I can't help the reflex action of trying to look more directly. It's to do with the arrangement of the rods and the cones, the light receptors, in our eyes. Cones are active in higher levels of light, i.e. daylight. They detect the different wavelengths of colour and are masters of spatial awareness – they are grouped together in the centre of our retina. Cones are what we see the world sharply with, so, of course, we instinctively move our eyes to look straight at something. The rods are more sensitive, they are the low light photoreceptors – they don't bother with colour and they are less good at spatial awareness. They are situated round the edges of the central conglomeration of cones, which explains my disappearing dogs. Counter-intuitively, though, we have a whopping 100 million rods in each eye but only 6 million cones.

So, when I describe what I can see I'm not really describing what's there – just what my eye/brain combo is making sense of. And now I come to think of it, it's another reason why drawing in the dark is so difficult: I have a blind spot right in the centre of my vision at night!

25 MAR | WAXING GIBBOUS
Daniel delivered some art supplies this morning. He also gave me some samples of the chalk pastels they are now making in the art shop. Lovely little parcels of scales of chromium oxide, burnt sienna, Prussian blue and vermillion. Vermillion! The darkest one in the box is the exact colour of a moon rising through the atmosphere into darkness. Why has it taken me so long to find vermillion?

28 MAR | **FULL MOON**
Full moon rise at 19:00 in the E. Perfectly aligned with the stream. Dark vermillion (!) moon, darker red reflection. The reflection is very long and wobbly. It seems to occupy the whole length of the stream. This is the spring equinox full moon and it rises almost due E. Perfect for viewing in the stream. The autumn equinox full moon will rise in the same place, too, but I won't be able to see it from here because there will be leaves on the trees.

Why do I relish the moon's reflection so much? I walk up and down the stream trying to find places I might catch it unexpectedly in. Is it just because a reflection is a doubling of the miraculous light? Maybe it's just because it's pretty? Is it light in the dark that makes it special? Or perhaps it's because the moon and water are in some ways twinned? There's obviously the direct link with the ocean tides but perhaps it's because water is as changeable, slippery and beautiful as the moon. The moon in the sky and the moon in water are both unattainable – untouchable – other.

I've been reading lots of myths and stories about the moon's reflection in water. The idea of drinking the moon is common in stories all over the world. So is the belief that capturing the magical healing properties of the moon in water can cure any number of ailments, including death. I flicked past a website the other day that suggested leaving a silver jug full of milk and rose petals outside on the night of a full moon. Drinking the milk the next day will rejuvenate skin. And if done often enough it will turn back the signs of ageing. Perhaps I'll try it, what the hell, it sounds quite delicious.

Suggestions for names for the March full moon: toad, frog, wind, rising-from-the-stream.

29 MAR | WANING GIBBOUS
The moon rose at 20:30. One and a half hours later than last night and much further to the S. Each moonrise at the moment is about an hour and a half later every day – but the moon is setting at almost the same time each morning?? Tonight is still and

cloudless. The moon arrived colossal and pink but is shrinking and turning gold as it climbs higher into the sky. I know that it doesn't actually change size and that the apparent difference is all in my head. When the moon is on or just above the horizon, my Earthling brain, intimate and entangled with Earthly things, incorporates it into the me-and-my-surroundings relationship. The thrill of seeing the moon in close association with known objects is heady and made doubly so because it's only temporary, as it pulls relentlessly upwards and becomes somehow unknowable.

Stars. Satellite heading east. Smell of daffodils and damp. Sound of toads.

4 APR | THIRD QUARTER

Easter Day is always on the first Sunday after the full moon that occurs on or after the spring equinox. In Christian Europe, that month was traditionally known as the paschal lunar month. The media call it the Worm or the Pink Moon now.

The Book of Common Prayer contains mindboggling tables of numbers and asterisked numbers for calculating Holy Days. Instructions include: 'To find Easter Day, look in the first column of the Calendar, between the twentieth Day of March and the nineteenth Day of April, for the Golden Number of the Year, against which stands the Day of the Paschal Full Moon. Then look in the third column for the Sunday Letter next after the Day of the Full Moon; and the Day of the Month standing against that Sunday Letter is Easter Day. If the Full Moon happen upon a Sunday, then (according to the first rule) the next Sunday after is Easter Day. To find the Golden Number, or Prime, add 1 to the Year of our Lord, and then divide by 19; the Remainder, if any, is the Golden Number; but if nothing remain, then 19 is the Golden Number.' This goes some way to making me feel less stupid about my moon observations… Big respect to all the priests in out-of-the-way rural parishes 400 years ago who had the brains to work out exactly when to celebrate Easter.

> 'The moon is born every month, increases, is perfected, diminishes, is consumed, is renewed. As in the moon every month, so in resurrection once for all time.'
> —St Augustine, Sermo CCCLXI: De Resurrectione mortuorum

> 'Jesus Christ … Was crucified, dead, and buried; He descended into hell; The third day he rose again from the dead; He ascended into heaven; And sitteth on the right hand of God the Father Almighty; From thence he shall come to judge the quick and the dead…'
> —Apostles Creed

It seems that the moon is at the very centre of this. Death and rebirth is a common theme in stories, myths and religions all over the world.

5 APR | WANING CRESCENT
Drawing the sun set from about 18:00 to 19:30. I draw it from the track round the top of the big field. The blackthorn blossom changes colour – from a shady white to a definite purple to a very deep violet purple. The sun changes from blue to orange to red. It seems to get bigger and then smaller as it sinks and heads W. Birds sing. I would happily do this forever.

10 APR | WANING CRESCENT
Sirius is bright in the SW. Its reflection quivers in the stream by my studio. They are intimately entwined, the heavens and this little river.

11 APR | NEW MOON
I haven't seen the moon for days and days.

Under the trees there isn't a glimmer of light; it is totally dark. I can't make out the definition of anything at all. The ground, which I know I'm standing on, the branches and leaves, which I know are sweeping down around me, have disappeared entirely; they have been replaced by grey sludge, and so has the space between me and them. It's the rough blurriness of night and it's right up against my eyeballs. Darkness appears to be thick and it's moving. This is called Eigengrau or 'intrinsic grey', 'dark light' or sometimes 'brain grey'. It's my optic nerve playing tricks on my brain, which apparently it does all the time but in better light conditions is easily cancelled out. It's visual noise, a bit like the background hum of a fridge that we can ignore. With my eyes wide open I stare and stare into the grey and I wonder if darkness is actually a thing. Does it exist? Is it just the point where human eyes fail to be of any use? Is darkness all in my mind?

When I stumble out from under the trees, they coalesce into shapes. The buildings, the ground, they all become distinct. I can now see a rainbow of blacks – and what blacks they are! There are bright blacks, bottomless blacks, matt blacks, blue blacks, textured blacks and white blacks. I can see them now because of the light – the light from the thousands of distant spring stars above me.

14 APR | WAXING CRESCENT
Low in the NW Earthshine. I'm taken by surprise by the lingering light in the north. Green sky. Definitely got a summery feel to the evening, despite being bloody freezing.

15 APR | WAXING CRESCENT
20:45 Moz and I walk to the big field and watch the moon emerge from behind clouds. It's still light enough to see Moz clearly. Complete silence except for some invisible and very quiet wing beats above me. Stars. Cannot

tear myself away. I think about the old moon in the embrace of the new moon.

I'm getting to enjoy the sky busy with so much more than just the moon. For some reason it used to worry me that the moon was often only a side show – that my subject matter wasn't centre stage. The night sky is so rich that a perfect crescent moon is a very small part of it. It eloquently shows its, and therefore our, tiny place in the much bigger picture. What I'm having trouble with now is: how do I show the moon's place in the 360-degree vastness? I feel the urge to draw everything – in front of me, behind me, above me, beneath my feet… I want to be able to show the moon in its place.

Because it's a crescent moon, and it's not creating too much light, Taurus is visible to its left. Aldebaran especially.

16 APR | WAXING CRESCENT

Midnight. Clear. Freezing. The moon has just disappeared behind the wood. The stars are sharp in the blackest black. In the patches of sky where there are no stars, like to the left of Polaris around the Lynx, the sky appears less black, greyer, smudgy. I'm sure this is to do with contrast. The contrast between bright star and darkness makes the darkness seem blacker. Where there is less contrast, my old friend Eigengrau creeps in.

19 APR | WAXING CRESCENT

20:30 Waxing crescent moon in Gemini. SE of Pollux and Castor. It's still cold but there are bats about. I disturb pigeons roosting in Luther's yard as I walk to the big field and arrive to hear the last of the birdsong and a very creaky crow. I've been drawing blackthorn all day and I've been revelling in its smell. It's not a delicious scent like honeysuckle or crab apple, but it's blossomy. Vegetative. A sort of basic flower scent. Perhaps the reason I like it so much is because I haven't smelled a flower since last summer. Blackthorn is the first of the year, and I'm just longing for a whiff of anything floral. It seems to smell stronger now, in twilight. And moonlight has transformed the blossom to snow; it's lying thickly on every black branch. There's laughter from the farm and a warbling tawny owl. Sirius is bright and a distant chainsaw starts up.

23:45 When I put my boots on to go out again, Moz, my right-hand woman, yawns extravagantly but comes with me. The moon is green and I'm surprised by a fly flying into my face. The tawny owl is doing its more traditional hoot.

21 APR | WAXING GIBBOUS

Moon waxing gibbous in Leo. I'm enjoying understanding where the moon will be tomorrow night and the night after. Because it's in a different place in the sky each night I can watch

above 21 April 2021

it travel through Taurus, Gemini, Cancer and Leo. I guess the sun is in the constellation behind the one you can see before it rises? I draw in the big field as the sun sets, through Civil twilight and into Nautical twilight. And then it gets too dark to see anything. With 4 bits of paper stuck to my board, which I'm holding flat, one edge pushed against my stomach, the small box of carefully chosen colours balances on the board. Nothing else. I look like the person who sells ice cream in the interval at the theatre as I wander around. It's exciting, not knowing what I'm going to draw, just opening my eyes wide and being ready to draw anything. Being out in the big field for the changeover of shifts between diurnal and nocturnal is joyful. There is the last and most beautiful birdsong from all around, from every bramble, gorse, thorn, wild rose and tree: the thrush and the blackbird's evensongs are spine-tingling. There is the cronk of the crow and the harsh bark of a roe deer. And finally there is the first quaver of the tawny owl and the scratchy sound of duck wing beats overhead. More blackthorn blossom this year than I ever remember. It's late but it's packing a real punch. White blossom against black stems is visually perfect in bright light, low light, twilight, and moonlight too.

22 APR | WAXING GIBBOUS
Went out looking for Lyrids, didn't see any.

24 APR | WAXING GIBBOUS
05:50 Sun rising ENE. Went out drawing. Heavy, frosty dew confounded me, it was too beautiful, too transient, the drawings were terrible. The moon rose in the E at 16:30. Waxing gibbous. 21:00 I am accompanied by the moon's reflection as I walk along the river. It ripples and splinters, comes together, shivers, expands, makes itself into

two, lets out jets of stars, gives itself a hat and then takes it off, dances and disappears behind branches. It stops when I stop and travels upstream as easily as it goes downstream. It encourages me into a mad quick-quick - slow - stop - run - about-turn sort of dance that I hope no one else sees.

25 APR | WAXING GIBBOUS
19:10 Thick grey cloud and cold wind. I get a glimpse of the almost full moon low in the SE. It's blushy pink and luminous, cosseted in an ocean of the softest grey. Weirdly it jolted me straight back to being about 8, groping about, looking for my poor hamster as it tried to sleep in its plastic hamster house. Coming across the little creature's surprising redness at the bottom of its billowing cotton-wool bed always made me start. Now I think about it, I'm sure I used to disturb it just for the colour kick it gave me.

26 APR | DAY BEFORE FULL MOON
20:10 The most amazingly vivid sunset tonight but I sat with my back to it, by the stream, and began to draw where I felt confident the moon would appear at any moment. Craning my neck, peering round the trees, but no moon. I was about to give up and go to the big field and draw the afterglow of the sun when I spotted the moon sidestepping me as always. Pale, pale pink and hazy behind clouds. I've got used to the moon evading me. It's partly down to this place: I'm always viewing it through trees. I've realised that being given the slip over and over again is what makes the moon irresistible. It's part of the fun. Initially I would dream of wide-open horizons, places where the moon would always be visible, places like Stonehenge or a clifftop with the sea on three sides. But now I don't think I'd swap being here, in this hopelessly wooded valley. I've got to know, or rather, I'm getting to know the moon locally. The moon here is a tease, a late comer and an early goer, a hide-and-seeker and a flasher. That the moon is a slippery customer is an understatement. For a start it's a different shape and in a different place every night. I'm beginning to think it behaves differently in every place it's viewed from.

Later, 21:30, with Moz on the big field, large clouds are stained orange by the moon and smaller, paler, squiggly ones are sharply in focus. SE is apricot with a small red light just above the horizon – I think it's the mast at Wymondham police station. There is a lot of light and noise. Sam's chainsaw. Clive knocking in fence posts. Someone shouting for a lost dog.

27 APR | **FULL MOON**
06:30 Drizzle, the moon has just set. Three fox cubs playing in the edge of the wood. 20:45 I'm in the big field in my favourite moon viewing spot. It's been raining all day (the first rain

in a month), the sun set half an hour ago, and the light is fading fast. I'm enveloped in gloom. My eyes are adjusted to the gloom. I probably won't be able to see the moon. Gentle dripping from the wood behind me. Quite depressing standing here really. Then 21:10 without any glowy fanfare the biggest, sizzling orange moon I've ever seen rises above the wood, in the SE. It's the silent appearance of a scalding orb into a murky night that does it – my heart races, my mouth goes dry, my knees really do tremble. I stare and stare with my mouth open, for what feels like hours, before I can attempt to draw. I feel I really should have fallen to my knees – but perhaps drawing is a form of worship. An unintended 'fuck me' came out of my mouth when I got the first glimpse: I wish my default surprised utterance was less basic, it's not much of a welcome for our incomparably exquisite, always surprising, unknowable, closest celestial neighbour. 15 minutes later the incandescent moon slides into the dark rainclouds and is extinguished. I feel totally exhilarated. Every atom in my body is vibrating. I slowly become aware of Moz panting heavily nearby. And I remember her setting off in pursuit of something the moment before I had seen the moon. I have to consciously think about my legs before I can move them, and walking feels unnatural. It's as if I'd actually left my body and now I'm having to reconnect with my flesh and bone.

Suggestions for names for the April full moon: blackthorn, dandelion, dry, whitethroat.

28 APR | WANING GIBBOUS

22:50 I'm walking up the road in the dark looking at the moon rise. No traffic, no sounds at all, except for the suckling of a calf on the other side of the hedge. The never-failing astonishment at the rise of the moon and the beginning of new life. There's a compelling juxtaposition between the dynamic visual spectacle and the absolute silence of a moon rising. In TV and film, the natural world is always accompanied by a soundtrack. Silence seems quite radical.

1 MAY | WANING GIBBOUS

Clear night. Moon not up till 01:00 or 02:00, so go out at 22:30 to try and draw the stars. Satellites like lice. Trundling along in every direction. Some bright, some much fainter. Shooting star high in SE. Crick in neck. Very difficult to look up and look down and then find where you were looking before, so try and draw without looking at the paper. Stupid really. But fun crouching on the ground outside my studio at midnight. Looking and looking and looking. Very cold. My breath obscures the sky when I look up, so I include my breath in my drawings. I can hear a swan beating its

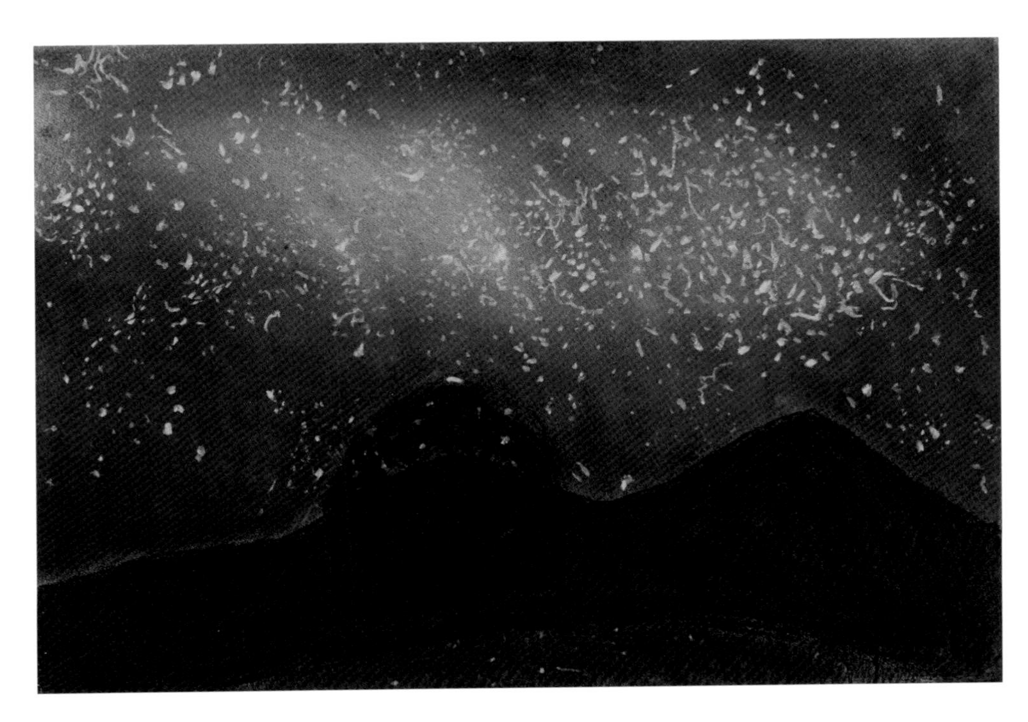

above 10 May 2021, Cumbria

wings at the otter I can hear whistling. Moorhens screech. Give up after a couple of hours because my neck is so sore. I'm fantasising about a dentist's chair contraption, with drawing board at a clever angle above me but not obscuring the sky. My feet are like ice and take all night to warm up.

10 MAY

There's no moon, it's nearly new. I'm in Cumbria. The heating has gone rogue and is on full blast at 02:30. Sweatily/blearily I go outside to the electric shed to turn it off. I stop dead – I don't understand what's happened. Something's wrong. I can't get my bearings. I don't know what I'm looking at. The sky is astonishingly bright. There is so much light, stars, colour, galaxies, infinity… I slowly realise that it's the Milky Way and I feel like I have been punched in the face by it. It's draped along the tops of Whiteside and Grasmoor and there are more stars than I've ever seen. The frosty grass glitters with starlight around my toes, and the usually incessant noise of lambs and calves and farm dogs is absent. The Milky Way is revealing itself and the world is struck dumb.

16 MAY | WAXING CRESCENT

Home. Another colour-of-the-moon thing. The dark irises outside the back door. The rivalry of rich dark petals and screaming cadmium middle is the same hair-raising combo as a moon rising in darkness.

17 MAY | WAXING CRESCENT
21:00 Thin waxing crescent. Cold. First sight of the moon for ages, it looks like a pale pink nail clipping lost on a dark, deep-pile, cloud carpet. We have had hundreds of swallows, martins and swifts here for about 3 weeks now. Sitting in trees chattering and endlessly swooping over the field and lake. We've never seen this many here before and feel uneasy. Why are they hanging around? Why aren't they busy nesting? Is this the only place in Norfolk with insects? As the light fades they disappear off to roost or float in the stratosphere or whatever they do, and bats take over. Hundreds and hundreds of big Daubenton's bats over the water and the field. They have almost the same flight pattern as the martins and swallows, but a bit flappier, and then a sudden jerk left, right, up or down. I'm giddy with this ceaseless movement.

21 MAY | WAXING GIBBOUS
Moon invisible behind thick grey cloud. Very cold, gale force winds and torrential rain tonight. I'm thinking about the two red kite chicks that we watched through binoculars, high in the ash tree on a heap of sticks. They only have the faintest white fuzz covering them. And what about the seven signets which hatched today. And every other tiny, naked, newborn scrap of life.

22 MAY | WAXING GIBBOUS
Coldest spring for years. Cold showers. Horrible to draw in.

23 MAY | WAXING GIBBOUS
Woken at 03:30 by 5 cockerels that have been dumped on the side of the road by the wood. It happens a couple of times every year. Does the person who leaves them here really think that the cockerels live happily ever after in the wild? What actually happens is that Sal catches them up and feeds them to her hawk.

25 MAY | DAY BEFORE FULL MOON
20:30–22:00 Cold. Thick cloud clearing from the NW. The sky busy and dramatic in the SE. There are sunset colours in the tops of the big cumulus clouds that are being blown apart fast. The May greens of the field have evened out and dulled down under the mayhem of movement and colour. Then I notice the moon, pale, still, poised in a clear band of sky beneath the blousey cumulus and a purple layer of cloud above the horizon.

The birdsong is incredible. There's a warbler in every bramble and thorn. There are song thrushes and blackbirds and so many I can't identify. I think there is a special bedtime crescendo which stops and starts and stops and starts and is more beautiful with each start. It's like being drenched in song. Every evening it sounds better than the one before. Every evening I think it would be impossible to surpass.

Ticks.

26 MAY | **FULL MOON**

Cold, thick cloud. I go to the big field at 21:45 not expecting to see much. The field is silent. It feels deathly. The sky is grey, it's dark. No sound. Of anything. Then boom…boom…… boom, boom. War-games begin on the Battle Area. And as if they have blown a hole in the grey lid, the sky begins to lighten in the W. Slowly, slowly the brightness spreads E. I see a star high in the S. Then a pink speck. A warm pink glow radiating through cloud and then a prick of bright rose gold and it's all over for the blanket of cloud. The May full moon is here. Immediately the flowering thorn bushes appear to float like icebergs through the dark vegetation. The scent of sweet rocket. I am very cold.

Suggestions for names for the May full moon: May blossom, white blossom, cow parsley, swifts, swallows, house martins.

30 MAY | WANING GIBBOUS

The moon rose at 00:53 this morning and it set at 08:21. At the time of the full moon it was rising over an hour later every night. Sometimes an hour and a half. And it was setting only half an hour later every morning. But now, only a week later and it's the other way round.

I walked to the highest point up the farm track at sunset. I wanted an uninterrupted view NW because I wanted to see Mercury, which is meant to be visible just after sunset. I hung around but couldn't make it out.

31 MAY | WANING GIBBOUS

I've just bought a Collins Book of the Weather. There's a section about phenomena which I'm relishing. It gives names and explanations to things that I have observed and drawn many times. For instance, Earth shadow – the blue purple belt of colour just above the E horizon, as, and just after, the sun sets. Above it there is often a band of pink called the Belt (or sometimes Girdle) of Venus. It's the place in the sky where Venus is visible (when it's showing itself) in the evening. The Belt of Venus is pink because of the 'backscatter' of red light from the setting sun. This all works the other way round in the morning.

On that first attempt, last year, to draw the January full moon rising I noticed and got in a muddle with Earth shadow and the Belt of Venus.

I've come to expect this colour configuration in the E sky at sunset, but I had no idea there were names for it. The Belt of Venus is such a deliciously poetic name. Venus, the Goddess of Love, her girdle, the heart-stopping pink – it all adds up to something to go out and look for. Naming the world adds layers of meaning to it.

23:00 The moon hasn't risen yet, it's a clear warm night. Instead of going to bed I can't resist going outside. I'm not going to draw: I just want my fix of wandering about, looking at this place in the dark. I think of nighttime vision as a superpower now. Why was I so obsessed with lighting the dark – with trying to 'see' in the dark? Why did I have such a narrow sense of seeing? I CAN see in the dark, and what I see is not the daytime world I mostly live in but a parallel one that's equally compelling. A modicum of patience is required to get there, but the reward is a slow unfolding of a place rich in subtlety and shadow where colours take on new meaning. It's a place given to sudden, astounding bursts of brilliance and wonder. My eyes can take me to this new world without leaving home.

I'm wearing my new silver flip-flops and they sparkle in the starlight. The dew is shockingly cool. Bats and satellites. And the incredibly bright International Space Station.

1 JUNE | WANING GIBBOUS

Clear, bright, warm morning. 09:45 I draw the nearly half moon above the poplars. The world is entirely green and white and blue. For once the moon is following the dress code; it's in faded white. In shade, the white blossom is blue.

I've been in contact with Dan from Great Ellingham Observatory. I had no idea there was an observatory at Ellingham. In fact, I had to double-check it was the same Ellingham as the one up the road. It's the peculiar building in the corner of the playing field. I always thought it was some sort of WW2 construction. He said I could come along tonight but I'm not sure what time 'tonight' is. I leave here at 21:15 and arrive at 21:30, it's still light, but Dan and Mick are just locking up and leaving. Whoops. They kindly open it up again and show me round. I'm apparently too late for solar viewing. Solar viewing! Solar viewing? I didn't know you could actually do that. The building is round and downstairs are loads of computers, telescopes, wires and other unidentifiable equipment. They are arranged on a curved work top around the edge and the walls are plastered with pictures of planets, stars, galaxies, black holes, etc. All the photographs are taken by members of the Breckland Astronomical Society. I feel like a newborn, my drawings so unscientific and clumsy, in comparison to this

Society of Astronomers. Upstairs is the biggest telescope I've ever seen. I would probably fit inside the main tube. When the roof opens and the floor revolves, I have to pinch myself that I am actually in Ellingham, with two really normal, nice blokes. They're just getting a perfectly run-of-the-mill telescope ready to look at some stars through – I'm not in Blofeld's lair about to be fed to piranhas. It's still too light really but I see Spica and Antares through the enormous telescope. Spica is 600 times brighter than the sun and lies on the ecliptic so it's often seen next to, or near, the moon and the planets. Antares is a red giant: if the sun was the size of a grape, Antares would apparently be the size of a small car!

Mick burst my idealistic picture of these august astro-photographers staying up all night getting good shots. Apparently you programme your telescope and then go to bed! WHAT? I spend more time outside looking than they do. Do they ever feel cold, or tread in cowpats? I realise that what I'm doing is not less. It's different. They've skipped my bit and jumped straight to the clever computer-aided stuff.

I ask Dan about the differences in the moon's rising and setting times. He takes a deep breath and he tells me it's all to do with the ecliptic. The ecliptic is the imaginary line in the sky of the path of the sun. Draw a line that follows the sun's path from the E horizon to the W horizon. You'll see that the line changes from winter to summer. Wider and higher in the summer, short and low in the winter. Twice a year that line crosses the Celestial Equator (another imaginary line in space; it follows the equator out into space and is visible from here because of the tilt of the Earth): those moments are the equinoxes. The moon and the planets also follow this ecliptic line – well, within 8 degrees of it. Dan says that when he's viewing the night sky he tilts his head with the direction of the ecliptic. What?? Then he starts writing numbers on the white board as an explanation. I really do try and keep up but he, like a rising moon, has got away from me before I can blink. It's 23:00 and my brain is long past taking anything in.

By the time I get home, the sky is magnificent. I try and tilt my head to the position of the ecliptic. I don't know what I'm doing so I just stand under the whole starry dome and drink it in. There's red Antares, the car to our grape, actually looking like a pimple from here. An owl hoots, a roe deer barks, and a moorhen shrieks blue murder.

2 JUNE | THIRD QUARTER

06:30 Last quarter moon at its midpoint in the S. A warbler sings its heart out in a young oak growing out of brambles. Strong song, strong growth, strong yellow green, pale sky with a paler moon above a horizontal diffuse vapour trail. There's a long comma of a cloud below the vapour trail and a thick squiggle above the moon. Like hieroglyphs. An unreadable code in the sky. Or maybe not unreadable – it says warm settled weather and a last quarter moon, all is well – and the warbler says 'IT'S JUNE'.

- The first quarter moon rises, roughly, in the middle of the day and sets, roughly, in the middle of the night.
- The third quarter moon rises, roughly, in the middle of the night and sets, roughly, in the middle of the day.

3 JUNE | WANING CRESCENT

Mary asked me if I'd been watching the golden moon. No. It rises so late I've been asleep. She has been waking at 3 or 4 am and looking out of the window. She says it's been a beautiful dark gold and very low in the sky. I'm jealous. I so wish I didn't have to sleep.

I wonder if you can work out the angle of the ecliptic for any given day? And if I could represent it on one of my moon charts? I ask Google, and the explanation starts with the equation $\partial = -23.45° \times \cos(^{360}/_{365} \times (d + 10))$. Ummmm, I think the answer's no!

8 JUNE | WANING CRESCENT

Nearly new moon. 20:30 I'm drawing a thorn bush, completely smothered in blossom that's turning pink as the blossom fades. The foreground is pink with ragged robin and there's a pink cloud above it. When the sun finally sinks below the horizon behind me, the blossom gets more pink. This surfeit of summer pink is like a perpetually overflowing jar of honey. It's too precious to be squandered but there's too much of it, all at once; how could I ever manage to scoop it all up?

23:00 It's only just getting dark. Yellow in the NNW. Oh hello, large, wet, keenly attentive mosquitoes – I'm not at all pleased to see you.

It's warm under the trees and cool in the open field. Ear-piercing squeaking from the lake.

9 JUNE | WANING CRESCENT

Big poufs of wild roses at the far end of the big field smell delicious. Mightier versions of my grandmother's swimming hat. They are the only plants on the field that don't have a browse line.

10 JUNE | NEW MOON

Usually the darkest part of the month but not now, not in midsummer. It's never dark. There will be a partial solar eclipse this morning. At 10:00 the moon will pass in front of the sun. It will be partially visible from the UK – that is, from here we'll only be able to see a bit of the sun being eclipsed. This is a nice time of day to be doing celestial viewing – I'm fully awake. I find instructions on how to make a pinhole viewer with two bits of card. Trying to secure the piece of card, which will act as the screen, into the right place at the right angle is taking longer than I thought. It's getting windier and the bloody thing keeps blowing away. I thought this was meant to be easy. With 10 minutes to go I get it sorted and notice impenetrable grey clouds sliding in from the W. They snuff the sun out. I stand about wondering if I might be able to see it through the cloud. No, it's uniformly thick and dark. 30 minutes later Isabel sends me a picture of it, mid-eclipse, from Great Yarmouth – I had told her about it yesterday. Bloody hell. I'm despondent, it feels like another failure. 10 minutes after the finish of the eclipse, the cloud clears and June blue skies and sunshine resume.

Solar eclipses only ever happen when there's a new moon because that's when the moon is directly between us and the sun – it doesn't happen every month because they often don't line up perfectly. Eclipses are only ever visible from some places on Earth. As I'm beginning to learn, viewpoint is everything when looking at the heavens. With this eclipse, we in the UK were on the very edge of being able to see it – the full eclipse occurred over Arctic Canada. I wonder if they had better luck with the weather?

23:00 Very warm, the stars are still faint. It's still Nautical twilight. Only owls, bats, mosquitoes and the tap-tap-tap of John Daniels's potato irrigation.

13 JUNE | WAXING CRESCENT
21:20ish The moon is low in the N. It's just above the wood. It's completely missable – like the tiniest tear in the mouthwatering pink gold purple sky.

14 JUNE | WAXING CRESCENT
Horseflies. The remaining member of the unholy Blackwater trinity. Mosquito, tick and horsefly. At least they only operate in the daylight.

21:00 and 22:00 Dark, agitated clouds roll in from the W. Strong golden light from the setting sun in the NW. Eventually it's blotted out. The blackbirds sing louder between rumbles of thunder. Pale turquoise sky can be seen under the cloud, to the S. Tempestuous black clouds, blue lightning, thunder, birdsong and enchanted last light. Then heavy rain for half an hour. They didn't get it in Dereham.

15 JUNE | WAXING CRESCENT
21:45 Waxing crescent moon is silver. It's high in W. There's colour everywhere I look. Pink, purple, orange, grey blue. Green dragonfly. There are golden gnats dancing in low light. Smell of honeysuckle. Venus is low in NW. A grey cloud like a huge paintbrush stroke. Thrush sings, cow moos.

Later – bats and gnats. Pink gold moon with thin veils of pink cloud drifting past, making it smeary. I went outside with bare legs for less than a minute and now have 2 huge mosquito bites turning into itchy welts on my leg. Pretty, skin-exposing, wafty summer clothes are for other people. If you live here, you have to keep covered at all times.

16 JUNE

I'm in Cumbria. 21:30 I scramble up the bottom bit of Mellbreak as a blanket of cirrostratus clears from the N. Its undulating underside is highlighted in pink. Finally the sun is revealed above Loweswater lake: Moz and I are caught in Golden Syrup light. The bracken is electric green, it's pulsing with green-ness, and about 8 pipits pipit loudly around us, flitting from one tall unfurling stem to another. We – the mountain, Moz, the bracken, the pipits and me – are all bound together, briefly, in celestial syrup.

20 JUNE

21:00 I'm drawing by the beck. Sunlight pours over the tops of the trees and into the shadowy gully. A sort of sticky, chlorophyll-orange resin coats every leafy surface facing NW. An expanse of dark cloud rolls in. The midges are behind my glasses and in my ears, up my nose.

Sunset 21:53 On my way home I catch a glimpse of fire through Kirkgill Wood – one last blazing minute of sun before the killjoy cloud puts an end to it for good. There's an owlet calling. Perhaps it's the same one that was peering down at me from a branch over the beck when I went for a swim yesterday. Its huge, dark eyes staring down at me, taking the world in. My – less huge and less dark – eyes staring up at it, taking the world in.

I stop to sniff the honeysuckle by the door on my way in. Too late I realise it is smothered in midges. So many they have almost turned the flowers black. They rise as one and envelop me.

21 JUNE

Waxing gibbous moon. But I'm not interested in it at the moment. I'm determined to draw the summer solstice sun setting. I've tried various vantage points, but I think halfway up Mellbreak is the best. It's a very good workout, all this rushing uphill

opposite & above 21 June 2021, Cumbria

carrying a heavy bag of art stuff, late in the evening. It's quite cloudy and not looking promising. But slowly, slowly as the sun sinks some colour appears. Then suddenly there is the blood red sun – Scotland is degrees of blues, the band of sky containing the sun is coral. Then the sky becomes dark red and Scotland turns coral. Fangs Brow is washed in the glow. The sun swells and swells and then disappears below the horizon. An intense half an hour. I've done 4 drawings. I feel elated. I can feel the spin of the Earth.

22 JUNE

It's midnight and Lorton Fells, to the NE, are astral blue. The sort of blue you only see in the north. It's a sort of fragile, swoony, transparent blue. The skyline is pink, fading up to yellow green. It's still light. I just want to keep heading north. I've got light madness. I don't mean a mild madness. I mean a madness brought on by, and for, light.

Depending on the time of year and the latitude of the observer, twilight can be over in less than an hour or linger throughout the night. In Cumbria (54 degrees N), from 17 to 27 June, the sun sinks no lower than 12 degrees below the horizon, meaning Nautical twilight is the darkest part of the night.

23 JUNE | DAY BEFORE FULL MOON

Home again. Very tired. 22:00 The moon is sluggish at the moment; it seems to be moving through the sky slower than at other times. It took an age for me to see it this evening and it's still behind trees. It's a blob of marmalade that can't compete with everything else in front of me. The garden is bathed in the lovely, long afterglow of the sun. Tree trunks are vermillion, their leaves are scarlet, lettuces are luminous, and bean poles are polished bone. How do I draw this? Everything is in a heightened state of glow, nothing is jumping out, nothing standing back. My drawings just look dull, there's no tension. Defeated again. There's an otter squeaking in the stream just below my studio. I'm off to bed.

24 JUNE | **FULL MOON**

Full moon arrived late. I am almost too tired to wait around for her. 22:00 She heaves herself above the horizon. Lipgloss pink in a lilac sky.

Suggestions for names for the June full moon: grass, elder blossom, dog rose, northern bedstraw, light, mosquito.

27 JUNE | WANING GIBBOUS

Very warm. At 04:00 the moon is buttercup yellow. At 04:30 the moon is the pinky white of a wild rose. At 05:00 the moon is a dandelion clock head. The dewy field is full of thin black slugs heading down. They're beating a retreat from the top of the grass canopy before the sun reaches it.

28 JUNE | WANING GIBBOUS

I've completely missed the moon this month. The sun has been all-encompassing, bewitching and beautiful. Light and sun are the June things. The light on the Lorton Fells at midnight was astonishing, I'm still thinking about it. Summer - light - sun - madness. My body clock is being stretched to the limit by this hunger for light. An insatiable craving for all the levels of light means sleeping isn't get much of a look in. When I do go to bed, I fall asleep wishing I was still outside in the simmer dim. I sleep fitfully and wake early, wondering what I've missed. The day feels rushed, but I know I'm being sluggish. Every little chore is a mountain to climb. I'd like a nap and I'm really looking forward to bed, but then I sense the light beginning to solidify and hear the birdsong thicken; my brain begins to fizz and I ignore the internal countdown to sleep. It's a pragmatic addiction, though, based on the probability of a change in the weather tomorrow. And a certainty of the change of season in a few weeks.

6 JULY | WANING CRESCENT

21:45 No moon. A day of heavy showers and sunshine. A clear, warm evening. I'm drawing the light on the remaining clouds seen over the tops of trees in the near field. A column of gnats above my head. There are 20 or 30 swifts swooping about; they often come to feed here in the evening. The young tawny owl in Luther's pine trees sounds like a squeaky dog toy. It's quite insistent. The other night it settled very close to my bedroom window and squeaked all night.

Sam starts up his chainsaw and a large military aircraft flies low and slow in huge figures of 8 under pink and grey and orange and purple clouds.

7 JULY | WANING CRESCENT

Total human silence and very disappointing blue sunset. Smell of honeysuckle and lime.

10 JULY | NEW MOON

Sunset in pouring rain. The atmosphere sometimes squeezes the just-above-the-horizon sun into unexpected shapes. I once saw a square sun set over the marshes. It's called 'refraction' and it's what happens when light bends. It's a mundane name for what happened just now. Tonight the tangerine sun appeared to drip through violet clouds and puddle into an oval shape. Then suddenly it was a crimson heart – and then it was gone. The smell of lime blossom and sweet chestnut and the sound of crickets in the late twilight.

12 JULY | WAXING CRESCENT

Black clouds boiling all around all day yield to a pink sunset. All the different grass seed heads are pink – well, actually they're purples and blues and browns and rusts and greys and mauves and pinks – but en masse they're an overall pink. A pink that I'm finding impossible to draw under a pink sky tonight. There are swifts, there are mosquitoes. Honeysuckle overpowering. Pigeons cooing from every tree. One of Clive's cows has been bellowing for 24 hours.

16 JULY | WAXING CRESCENT

Midnight. Half moon just set in the W. Incredibly bright satellite SSE. Long red shooting star in the E. The Milky Way, through my new binoculars, looks like weathered rock. The bright stars are the points at the end of pinnacles. There are dark crevices and shadows wrinkled about beneath. I'll have forgotten it by the time I get to my studio tomorrow. Dolphinus. Very noisy tawny owl chick on the edge of the stream. Voices. Cold nose. Dew.

18 JULY | WAXING GIBBOUS

21:00 Waxing gibbous moon in a pink sky, low in SW. Steely blue sky to the E with snow-capped moustache-shaped clouds. Swifts over the wood tonight, rather than over the lake. There are layers of sound: a thrush, a whole field of crickets and grasshoppers, and then above it all the whine of gnats in the Roman cork oak. It smells dry. The smell of sweet chestnut blossom is everywhere. There is also the smell of hay.

19 JULY | WAXING GIBBOUS
I've been looking at the moon rise and set times in Tromsø for this time of year. As always, I'm bamboozled. The new moon follows the sun so it makes sense that when the sun is above the horizon all day, the moon will be too. It doesn't set for a week. I get that. And the full moon is opposite the sun so it's below the horizon for about a week. But today, just after the last quarter, the moon rises at 19:00 and sets at 22:00 and only gets 1 degree above the horizon. Can you even see it? Probably not, and when the sun is up all day the moon is a side show anyway. The curve of rise and set times between the never-setting and never-rising is steep. If I was clever I could use it to help me work out rise and set times here. But I'm not.

23:00 Yarrow must be twinned with the moon because it comes into its own in strong moonlight. There's hissing from a swan on the lake and a softer hissing from the long grass as Moz moves through it. The odd grasshopper is still singing. It's a breathless evening and high cloud is spreading from NW. A looney in a loud car on the Hingham road takes the corners fast. The owlet calls continuously.

20 JULY | WAXING GIBBOUS
Crickets so loud. Crickets in the house. Crickets in the dogs' beds. Crickets on the dogs' heads. Behind the fridge. In my shoes. Crickets jump up and surprise me everywhere. Cricket crescendo. Long, long fingers of pink cloud brush over the rising moon. Hot. Moths. Tired.

opposite & above 21 July 2021

21 JULY | WAXING GIBBOUS

21:00 I walk up the path to the village to get a good view of the sunset. There's about a mile of wheat between me and the horizon. It's interesting to watch the light on this extensive single surface, which appears a sort of musty blue when the sun is above the far trees. It's quite dull and dark by comparison to the sky. It's easy to make it too dark in my drawings. I do it every time. The patches of wild oats, above the uniform wheat height, are pinker, lighter. As the sun sinks, colour slowly comes back into the wheat. When the sun is below the horizon, the landscape lightens, the wheat is dull pink and the wild oats are now the colour of sunset, the trees are black and the sky is yellow.

When I turn round to walk back home, there is the rising, nearly full, moon in front of me. It's glowing, a sort of cool pink, in a violet sky. The wheat, trees and hedges are all vermillion. 180 degrees difference and I could be in a different world. I turn NW, back towards where the sun was. I turn SE, back to the moon. I'm standing in one place, on the same spot, but the world appears to be two different things.

The dogs are with me and as we round the corner we surprise the neighbour's cat. It dashes on to, and along, the road – followed by my two lurchers. I can hear a large tractor and trailer approaching. Oh God. Amazingly the cat gets away and no one is squashed. It brings me back to earth with a bump. Harvest has begun. Noise of agricultural machinery in every direction.

23:30 The moon is not above the trees yet. It's big and bright. Clear, warm night. Am too tired to stay up. I feel defeated and that I should be keeping vigil with the moon. I have a tussle in my head: sun - moon? bed - moon? In the end, it's the all-consuming summer sun and the urge to grab the light that wins. I go to bed.

22 JULY | WAXING GIBBOUS
Went drawing on the track to the village again. The pale moon gargantuan next to telegraph wires. I put the dogs on leads tonight. One lead in each pocket, so my hands are free for drawing. We all get in a complete tangle. What a pain in the neck. If they'd seen the cat, I think they would have torn my trousers off.

24 JULY | **FULL MOON**
Full moon rising unbelievably late. Not due till 22:00 so I finally see it at 22:30. Why is it later than the June full moon? I watch it rise from the edge of the pit hole. Bigger than the sweet chestnut trees on the Hingham road. From fresh to smoked salmon, bowling along the main road, too heavy to leave the ground. Silhouetting each tree in turn. It gets all the way to Kevin's house before it finally takes off. I only did one drawing. I can't believe this has been going on all my life and until now I've never bothered to turn up to watch it. Elated.

Suggestions for names for the July full moon: docks, grasshoppers, crickets.

26 JULY | WANING GIBBOUS
22:40 Warm, quite cloudy. A combine is still going. Its noise gobbles up the mystery and vastness of the dark. Helicopters endlessly. I wonder if they are actually looking for something. The Summer Triangle – Deneb, Vega and Altair – seems so obvious in the sky that I can't really believe I couldn't find it last year. It's nicely dark at 22:30 now. Nights are getting longer. 02:30 Strong golden light from a slightly fuzzy moon, minus its leading edge, high in the S, in the early hours.

27 JULY | WANING GIBBOUS
20:55 Drawing the sun set from the track again. When the sun was 9/10 below the horizon, a very pure pink beam of light appeared to come from it, pointing straight upwards. The beam was as wide as the sun. It turned the burnt sky and the purple cloud pink. It ended in an extraordinarily bright glow above the cloud. Almost like a false sun. How amazing. How unexpected. I'm sure this has a name. It was like a pillar of pure light.

00:45 I see a vivid, long-lasting, emerald green shooting star in the W. I stay outside for another hour in the hope of seeing another one. I don't. Knackered.

31 JULY | THIRD QUARTER

Last quarter moon is high in morning sky. 22:30 Warm, cloudy and still. Silence except for the sluice after yesterday's heavy rain. Capella is rising in the NE. Huge and fuzzy, bright and crackling, changing from red to orange to green, when seen with bare eyes. A smaller red orange green squirmy thing, when seen with my glasses on. A tiny pinprick traffic-light speck that won't keep still, when seen through my binoculars held in my shaky hands.

The last thing I do before bed is stick my head out of my bedroom window. I do it every night now. I can't leave the night sky alone. The cloud has broken up and the stars are uncountable. I'm too tired to go out, I'm on the downward slide into sleep. I end my day with wonder. I begin my sleep with wonder. Either way it's good.

1 AUG | WANING CRESCENT

14:00 Faded moon setting in the NW. We've moved the buffalo from the big field to the far field and I've been seeing a roe doe and her twins on the big field. One twin is always next to her and jumps when she jumps; it shadows her. The other is always further away from them, doesn't notice its mother notice me, and often runs in the opposite direction from them. I know she won't tolerate my regular presence on the field; she'll move her family somewhere quieter. The buffalo bar me from this field, which allows others to move in and make it their undisturbed home. The movement of buffalo and human zones across three fields makes me acutely aware of the impact my dogs and I have on others.

2 AUG | WANING CRESCENT

Waning crescent rising NE. 00:30 The moon is a pale tangerine segment crowning the game covert. Darkness and this glowing, fleshy piece of fruit balanced on top of the trees has changed the meaning of this view. The small tightly packed square of fir trees is no longer a shooting-induced blot on the landscape but a bejewelled crown. The sky immediately around the moon – and especially around the leading, distorted terminator line – is midnight blue. Silence.

3 AUG | WANING CRESCENT

08:30 Same moon but half the size as last night. So high and pale. I love the fact that my brain can alter the size of things, and by doing so give me so much pleasure. Last night's moon seems like a dream. I have to tip my head right back to see it this morning. Above me in the blue sky is the white moon; between me and the moon are 2 long white contrails, a few fluffy white clouds, 3 white gulls, 10 white undersides of swallows' bodies, lots and lots of white thistle down.

Tuesday night at the observatory. The other members hit me with a dazzling array of facts.

- In the centre of Jupiter is a diamond the size of Africa.
- Solar wind isn't wind.
- The sun has two north poles and two south poles.
- Satellites going E/W are generally weather satellites; ones going N/S or S/N are military or spy satellites.
- Depending on what they're made of, meteors burn different colours as they enter Earth's atmosphere; green meteorites are probably caused by oxygen burning.
- Something fascinating about infrared, which I've forgotten already.

In fact, I feel like I've been bombarded by meteorites. I wish I could remember everything.

Get home and take the dogs into the field. Silence except for the young tawny owl and the sound of dogs hunting rabbits in the long, dry grass. The Milky Way is fathomless, and satellites abound.

4 AUG | WANING CRESCENT

Waning crescent moon is just ahead of the sun this morning – just out of reach of its rays. Just on the edge of where the sky turns blue. Very high.

21:00 Drawing dusk by the gate into the big field. A lot of farming noise. Rain is forecast. The sound of revving and revving and backfiring. Crows. The young tawny owl is screeching with great urgency just behind me. An adult calls. It calls again – nearer this time. And then nearer still. They are both right behind me. I mean literally right behind me. I can't resist, I turn my head to try and see them, and just catch sight of them flying away. Silence. Damn.

22:00 The farm machinery is finally silent. Stars begin to appear. Jupiter. A very bright satellite, heading S, fades and disappears. Bats.

6 AUG | WANING CRESCENT

Blustery day of intense sunlight and thunderstorms. Squashed toads on the road. A motionless sunset. For about 20 minutes everything in the lingering light gets impossibly pink and then pinker and pinker while the shadows get darker. The sky lingers yellow. In the E there are the same ice-white moustache clouds in a blue grey sky that I saw before. Grasshoppers still chirrup. A small breeze starts up again. Roe deer chase each other through the wheat; it makes a dry rattle. A hare. The poplars hold on to the pink afterglow for ages. I stop on the track to draw them: the trunks are puce. Hornets pass me, returning to the wood.

There comes a certain point (at the final fading of the pink light? Is that the end of Civil twilight?) when my glasses are no longer a help with seeing. Suddenly, and unannounced, a bright gold fireball with a slightly wiggly tail slips in from the E. It seems to explode in front of the low ice-white moustache cloud that I'm looking at. Just like a comic book KERPOW. What?! I've never seen one so bright and so apparently near. In (almost) daylight. Is it a firework? But there's no noise. I wait to see if there are more. No. I don't know what to do. I feel I should do something… I'm astonished. I desperately want to see it again. And again. Press replay. Really concentrate on it. But there's nothing – just a run-of-the-mill August twilight in mid-Norfolk.

10 AUG | JUST AFTER NEW MOON

The fireball has been on my mind. I report it to the UK Meteor Network. Why? Perhaps because no one here has been that interested. 'Oh' is the best reaction I've managed to elicit. I'm sure the UK Meteor Network are a band of like-minded folk who know the thrill of a good fireball when they hear of one. I've seen so many shooting stars since I started watching the sky, I'm beginning to understand that they are literally raining down on us. It's a process as fundamental and humdrum as rain or falling leaves. I just haven't been aware of this heavenly dandruff before. I'd always thought space was empty and up there, that it was a separate realm. What I'm beginning to understand is that it's teeming with stuff, and that that stuff is everywhere. Including here.

Golden light just before sunset, but I'm cooking. Pigeons turn the thick syrupy light into sound. These sweet summer evenings with the heady cooing of pigeons are paradise.

20:15 A perfect, warm, still, clear evening. Walked towards the village with the dogs to watch the sun sink behind Edward's fields. There's a sliver of orange gold moon low in the W. On our way back, two huge tractors with long trailers and yellow flashing lights charge past us on the road. I doubt they even see us. They are terrifying. Outsized. Bigger than houses. Lights blinding. Chewing up verge and toad and overhanging branch and anything else that lies in their path.

Twilight. 21:00 In the big field. The ragwort is still obviously yellow and mist is hanging over the boggy ground making fantastical shapes. The moon is getting bigger as it sinks lower. It's a fiery sickle of humungous proportions. I walk into the mist expecting to feel a temperature difference but there isn't one and the mist seems to vanish as I arrive. As I move, it moves. Shapes are distorted. Shapes are magnified. Shapes are hidden. Shapes are half-hidden. Thorn and ragwort and mist and thistle down and me – entangled, floating, solidifying, vaporising.

I notice my heart beating. I try and draw, but I'm not sure what I'm looking at. This is one of those altered reality moments of complete extraordinariness. Without any fanfare you're just suddenly in it, and then something changes – in this case, I think the light level dimmed – and with no warning it's all over. And you can never go back. Ducks are scurrying overhead in every direction.

11 AUG | WAXING CRESCENT

20:45 One or two clouds but a mainly clear sky. I draw the last light on the poplars again (secretly hoping for another fireball). The air is hot still, and I'm even hotter because I'm wearing wellies (to protect against ticks) and my coat (to protect against any other biting insect). My latex gloves (which are horrible but which I always wear now to keep the poisonous chalk pigments away from my skin) are filling up with sweat. I am a volcano radiating heat. At this time of day I get the last of the horseflies and the first of the mosquitoes. I hear Moz snapping at the mosquitoes.

The Perseids peak tomorrow or the next day. I seize a clear night and at 22:30 take my big studio chair out into the field and settle down. I'm early, I know it's best after midnight, but there's only so much time wasting I can do at this time of the day that doesn't bring on sleep: watching telly = sleep, reading = sleep. Gradually I become aware of the dry sound of birch leaves stirring, next comes the clatter of poplar leaves and then, as if organised by some horrible party pooper, thick clouds blow in from the W and extinguish the stars one by one.

This is what I saw moving above me tonight:

- Perseids – 1 (faint)
- Satellites – 5
- Bats – lots
- Gnats – lots

13 AUG | WAXING CRESCENT

The sky has been full of every sort of cloud at every level, streaming through on a strong, warm SW breeze all day. Sunlight strong and very glare-y. The ragwort is almost too yellow to comprehend. The end of the big field and the far field are almost completely yellow. This year has certainly been a good year for it. It's about shoulder-height and the buffalo seem to swim through it; only their heads are visible. At dusk it's still punchy. Well into Nautical twilight I can't see to draw anymore, but the ragwort remains yellow. The bumpy bottom of the altocumulus cloud is stained a cream of tomato soup colour that I can't reproduce. Grasshoppers still sing, crows chatter. Sultry.

16 AUG | WAXING GIBBOUS

The wind was in NE this morning. It's in the NW this evening and it's noticeably cool. I'm wearing my disgusting, but warm, drawing coat. Longer nights, filthy old coat: it means only one thing – the end of summer is nigh. Drawing ragwort in the dusk again and I'm dismayed that I seem to have lost the moon. I know where it should be, but I can't see it. I'm embarrassed. Then suddenly it surprises me much lower than I was expecting. Its light seems to jump out from behind stubborn clouds. It gives me a jolt. 'Boo,' it says, 'don't get complacent'. It's the same shape and only a bit brighter than the ragwort heads I'm drawing.

The grain dryer is on. Gnats are dancing in a column high above my head. Am I creating a thermal? An updraft? I actually see an owl tonight. Just. And I imagine how much better she can see me. Bats. Moths. Muntjac.

17 AUG | WAXING GIBBOUS
Thick cloud. NW wind. Rain. Dark. What a relief to be able to go to bed early without feeling I'm missing out on anything. The feeling that I must be out drawing/looking all the time, the every-moment-counts madness of light for 24 hours a day, is easing. We're tipping into darkness and I feel I don't have to be out drawing all day and all night. Phew.

19 AUG | WAXING GIBBOUS
22:30 The moon is appearing and disappearing behind a sequence of intricate, unfolding, dark cutout shapes. It radiates pink and red and gold and yellow onto invisible phantoms, behind the cutouts, who take the light and play with it. They pool it over here for a moment and then bend it over there. They scatter it towards the horizon or they lob it up to heaven. They can extinguish it, too. Everything is fast-paced up there but there's no breeze at ground level. No noise anywhere, except Moz who is pouncing on voles in the dry grass. Out of the corner of my eye I can see the white tip of her tail wagging. Poor voles. Jupiter is bright in the SE.

A large cloud low on the northern horizon is slightly orange, or is it my eyes? As I look at it, its colour seems to swell. Redder and redder. The strip of sky beneath it is an exquisite turquoise. And then, imperceptibly at first, like a slow exhale, the colour fades.

Later there are otters squeaking and moorhens shrieking.

20 AUG | WAXING GIBBOUS
20:45 I walk up to the crossroads. Billy's coming home tonight, I hope I might meet him there. It's very warm. A silent sparrowhawk flies ahead of me. A heron flies next to me, making earsplitting heron noises. Pale verges. Pale earth. Pale wheat. Pale thistle heads. Crickets. The moon is apricot and appears to be confined inside a pale pink box. Slowly the apricot ember ignites to molten iron and the edges

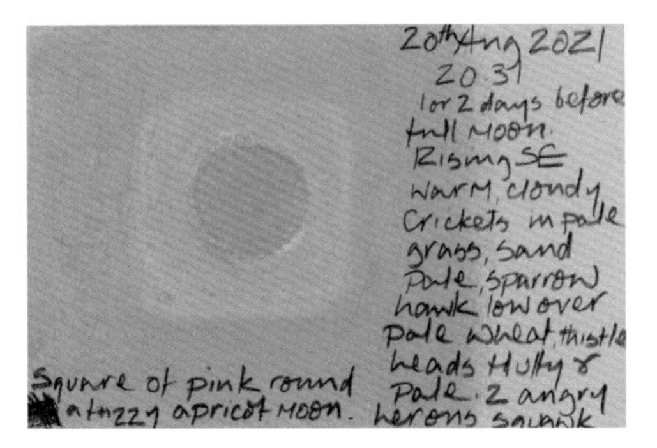

of the box dissolve and darken as the inside burns vermillion.

Much later I'm on the big field. I'm drawing. Moz settles down in the moonlight. She curls her body into rushes, head up, watching, listening, nose twitching. Back in my studio I see that my night drawing system – taking one small box with a few carefully chosen colours – has failed. Instead of getting it ready in daylight, I scrabbled about gathering everything in the low light from the useless bulb above my desk. The blue I brought was green, and the vermillion was brown. Funnily enough, although it's a slightly odd drawing, it's very obviously what I was looking at. I suspect I'm spending far too much time worrying about exact colour when it's the relationships between colours that's important. Still putting the moon in the centre of the paper. It's annoying. It's unconscious. I think it's because I'm drawing fast – too fast to think!

21 AUG | DAY BEFORE FULL MOON

20:30 The moon rose behind a thick front of cloud and is just beginning to make itself noticed. An astonishing sight – dark above and dark below – the moon is illuminating an intricate, lace-y mother-of-pearl fringe along the edge of the disappearing front. A heavenly version of the collar in Van Dyck's 'Charles I in Three Positions'. An opalescent hologram draped atop the jumble of thorns on the other side of the stream. I walk into the field and watch the moon free itself from the filigree. Mist begins to rise. I crunch 3 snails underfoot on the step outside my studio – I hate doing that.

The smell of crab apples is strong.

22 AUG | **FULL MOON**

20:30 I'm in the big field waiting for the moon. It's a clear evening. There's mist in the bottom of the field. Jupiter and Saturn are visible. The ragwort is still yellow. Crows, ducks, gulls on Edward's newly drilled field. And then a blanket of cloud spreads rapidly from the E. Bugger. Twiddle my thumbs for a bit. Mess about taking selfies with Moz. Twiddle my thumbs some more and am about to go in when a pink fuzz appears low in the E. This month's full moon is rising more to the east than the July full moon and more to the east than the moon of two days ago. For about ten minutes, bits of a blisteringly hot full moon can be seen – in and out of clouds and lighting up tiny areas of

above 22 August 2021

cloud, in smudges very close to it. Then the cloud thickens and it's completely gone. Everything is grey. Everything. Even the ragwort. But the crickets are still singing.

Suggestions for names for the August full moon: cricket, ragwort, cooing pigeon.

23 AUG | WANING GIBBOUS

05:00 Clear sky, twilight, no stars, Jupiter not visible. The setting full moon is low in the W, it's red gold with a defined dark outer rim. Complete silence. The owl that has been hooting all night is silent. The apparent opposites of silence and spectacle make it unfathomable. The sun is always accompanied by sound, traffic, voices, birds, etc. Then a wren strikes up for several minutes before a heron joins it to bid farewell to the August full moon – an avian salute – the biggest and the smallest bidding farewell to the August full moon.

24 AUG | WANING GIBBOUS

More clear than cloudy. This succulent, horizon-hugging, lopsided August moon is rising only a few minutes later each night at the moment – although it's rising much further to the E each night (don't ask me why). I decide to go to the big field, to my favourite summer moon watching place.

Pigeons and tawny owls call, ducks fly fast overhead. A muntjac barks. Then, crashing from brambles in front of me. Oh God, are the buffalo in this field? Have they broken out of the far field? I didn't check. For the first time I am properly alarmed to be in the dark. There's more crashing and Moz is staring intently towards the noise. I'm getting worried. Then coughing. Coughing? I've never heard the buffalo cough. If it's not the buffalo, then what is it? Crash, crash, crash cough... then, much to my relief, I make out the shape of a roe buck. I feel myself relax but almost instantly tense up again as a thought occurs to me: I wonder if the roe deer is coughing at the buffalo? Will I see the black buffalo in the dark? What will they do when they see me? Darkness and very large, unpredictable, horned animals have hit some ancient nerve. An uncoiling of terror, neatly folded up and packed away generations ago, has sprung to life. This doesn't feel like my safe, familiar Norfolk field in the 21st century, this feels prehistoric. Eventually there is silence except for a pigeon and an owl. My drawing is jittery, though, and I find the bird noises irritating.

26 AUG | WANING GIBBOUS
I found the freshly dead stoat on the track on the 22nd, full moon. Today it is almost all gone. I've been watching it decompose, although decompose is the wrong word – decompose sounds sedate, like a slow fading away – when in fact the stoat is being voraciously devoured. The devouring is so loud I've been able to hear it when I crouch down to look. First came the greenbottles, then the devil's coach horses and the undertaker beetles. The eyes were the first obvious thing to go. Then the undressing began: a ring of hair is suddenly evenly spread around it. The next day the chomping could be heard. The little body, originally lying on its right side, was now spreadeagled on its front, and if I looked closely it was moving. Undertaker beetles occasionally appeared from underneath and disappeared back. Devil's coach horse beetles came out in the open more, curling their tails into the air. Today the stoat is flat, there are 3 large visible exit holes, the tail has been snapped off, and there's a lot more hair lying round it. No undertakers or coach horses, just lots of maggots fleeing the scene. I'm amazed by how quick this disposal has been. In 4 days it's gone from stoat to a collection of black skin, bones, hair and feet. In a week or so the maggots will hatch into flies. Perhaps a young swallow will eat one, or maybe a dragonfly will, or a spider. The stoat will be in many places at once. A Blackwater stoat recycled and recycled. Perhaps a robin will eat the spider (that ate the fly), and perhaps a stoat will raid the robin's nest and eat its fledglings. Round and round. Dead on the full moon, a mass of pupating maggots before the third quarter moon, a swarm of flies by the new moon, being digested by the hobby (who ate the dragonfly) by the next first quarter moon. The circle of life, the circle of the moon, the connectedness of everything.

27 AUG | WANING GIBBOUS
An orangey moon rising at 22:00. It's clear and warm (ish). Tonight I'm not drawing. Unshackling myself from the delirium of it is a huge relief. No faffing about getting ready. No rushing. No hanging about waiting. No worrying about good viewpoints. No wishing the moon would slow down. Just wandering about in the dark, looking at the moon but also looking away. Going to places I know won't be good to draw from but are actually interesting to see it from. Watching the colour change without feeling I have to match it. And best of all, just turning my back and coming in when I feel like it. Several shooting stars.

29 AUG | WANING GIBBOUS
03:30 Pre-dawn moon high and still climbing. Orion is rising in the E.

It gives me a wintery jolt: I haven't seen Orion since April.

20:45 I stand outside my studio in the gloaming. Everything solid is dark. Moon not rising till 22:30 but dark enough for stars by 21:00. The Milky Way is clearly visible. Two military jets rumble and their bright lights constantly catch my eye. They are travelling fast and the movement confuses me: they are almost like shooting stars. They are also really, really bright. Competing with Capella for brightness low in the NE. No owls, no grasshoppers, just loud and ceaseless jet noise. Backwards and forwards, backwards and forwards. The noise changes and spoils everything: the silence obviously, the feeling of solitude, the infinity of space, the comfort of darkness and even the rich smell of plums by the gate is diminished by it.

When the dark orange moon silently rises in the NE, the jets are made insignificant by it.

31 AUG | WANING CRESCENT

Hot, still night. I decide to sleep outside. I lie down looking up S at Aquila. Altair is v bright. Jupiter below. It's lovely to be lying down looking at the stars. Drifting off. Waking up and drifting off. At some point the moon appears above the house, in the E. Rustling in the roses. Wake to find the stars gone. Mist obliterating the bottoms of the trees; the moon, now in the S, the only visible celestial body. Moon moves through the sky as I sleep and wake. Heavy dewy mist on everything, but I'm very cosy in the bed. Tawny owls in Luther's yard call to each other as night ends. A robin is the first daytime bird to make a noise. And then a pigeon. I think I've never heard anything so beautiful. Eventually get up to make a cup of tea. It's 05:00. Later when I walk round the field with the dogs I can look straight at the sun through the mist. To me it's a moon-sized pale disk just above the horizon. To the moon, which is well above the mist in clear blue sky, the sun's brilliance is burning its back edge pink.

2 SEPT | WANING CRESCENT

Cloud so thick it didn't seem to get light all day. 22:00 Night envelops, grey-dark, warm, soft, silent except for crickets under the Scots pines. This greyness in the dark is comfortably dull. It's what most English evenings are made of. Unremarkable only because they happen most nights in this warm, cloudy country. Greys gently dimming to darker greys as the circadian rhythms of all living things turn. And I can go to sleep without feeling I'm missing out on drawing opportunities.

3 SEPT | WANING CRESCENT

For a while now I've noticed that standing on my own in the field in the gloaming or in the dark, looking and looking, and listening to the absence of noise, gives me a physical sensation. It's a mixture of relief, or maybe release, like a long, unhurried exhalation combined with a sort of tingling, shivery feeling on the surface of my skin. It's as though I'm being swaddled. In daylight I have space around me. I like to think I am a solid free-standing creature but at night darkness invades that space, it bumps right up against me and crosses what I thought were my edges. It's deeply comforting and I feel wildly alive.

4 SEPT | WANING CRESCENT

22:00 The pigeons have been cooing so loudly all day that I can still hear them in my head. Like tinnitus but nicer. Thick cloud rolled in with twilight.

22:30 I'm standing in a blizzard of bats outside my studio. Coming from every direction, they miss my head by millimetres. I can hear clicking – I mainly hear it behind me. They make grotesque black shapes against the pale sky as they twist and turn and dive round my head. Then they vanish against the dark vegetation. The moths, though, are pale against the dark vegetation and disappear against the sky. The church bells suddenly peel out.

5 SEPT | WANING CRESCENT

Hundreds of swallows around at the moment. And a hobby. 23:00 Hot night. I'm sleeping outside again. The Great Square of Pegasus is rising in the E. This is an easy one to identify – it actually IS a great big square. The traffic on the Hingham road is ceaseless tonight. There seem to be a lot of motorbikes. Sleep on and off. Altair moves W. Something on the lake – maybe a moorhen? – begins to hiccup loudly. Over and over again. The motorbike noise is still loud. Why so many motorbikes? Is it getting lighter? A heron shrieks. I give up. I'm going to cut my losses and go inside in the hope of an hour or two of uninterrupted sleep. I walk past the bathroom window and something makes me stop in my tracks. I look out and see nothing but night. And then it catches my eye. The skinniest sickle moon I've ever seen. More skeletal than skinny, low, so low in the NE.

7 SEPT | NEW MOON

Very hot, cloudless day, lovely warm evening. Warm blackberries. Squirrels everywhere. Drawing the sunset in the E. It's amazing how quickly it has shifted along the horizon. The long, late, slow sunsets in July seem like minutes ago. I wouldn't be able to see it set from the village track now, so I'm back in the big field watching it sink behind trees. Wonderful fan of lines of pale clouds that look like I scribbled them. A combine harvester on the other side of the lakes. The tawny owl does its first bubbling hoot at the exact moment the sun disappears below the horizon.

15 minutes later the mosquitoes begin to harass me. Bats. Crows. Large groups of ducks scurry about. 2 rifle shots. The smell of blackberries.

8 SEPT | WAXING CRESCENT
06:45 Warm, misty, dewy morning in the big field. I can hear a tractor sprayer whooshing along the road. The dogs are soaked with dew. I'm drawing and the mist is tying me up in knots. I've been drawing day into night and often my pictures just get darker and darker. There's an urgency and a cut-off point. Drawing night into day has the same urgency, as everything is changing just as fast, but with the luxury of being able to see what I'm doing. There's no cut-off point this way round. Drawing fast is quite addictive. I have to make colour decisions and stick to them. Sometimes the decisions are weird, sometimes completely wrong, and at other times very exciting. It's an intoxicating feeling, tripping over the slowness of my eye-brain-hand coordination. It feels dangerous and out of control compared to the deliberate drawing in my studio the next day. A drawing in a couple of hours in steady light seems no challenge at all. I'm addicted to 5-minute drawings in precarious light. Am I ruining my attention span? Is a 5-minute drawing the same as scrolling through social media? Will I be able to execute a slow, considered drawing ever again?

Another hot day. 20:15 At Nautical twilight the sky is a phenomenal gradation of blues. From yellow blue at the horizon to the deepest lapis in the zenith via every blue imaginable.

22:00 The warm SE breeze has dropped and the sound of traffic is remarkably loud in the still air. Lots of shooting stars in the NE. One long, slow, gold one and one turquoise one. Maybe my eyes are doing funny things but I keep seeing sort of invisible shooting stars. Pale streaks, movement above my head? There are lots of satellites tonight and a very bright slow flash. Is it one of the out-of-control satellites spinning through space that someone at the Observatory told me about? As they spin, their mirrors catch the sun and they appear to flash. It sounds like there's a runaway traffic problem on the Hingham road tonight. The noise is intrusive. The Milky Way is strong in the zenith. Jupiter is very bright and Capella is twinkling. I remember standing here noticing Capella rising in the E this time last year. The crickets are loud under the trees. I stand on a spent giant puffball and give myself a fright. Horrible sinking feeling, as if one foot is being swallowed.

I hear Moz pounce in the dry grass and hear a vole squeak loudly – I momentarily hate her for it. So much loud traffic. Everything feels scratchy and off kilter tonight.

9 SEPT | WAXING CRESCENT

The sun sets at 19:23 tonight, the waxing crescent moon is close behind and sets at 20:38. Missed it as I was cooking, washing up, etc. Annoying. I go out later, though, to get my fix of the dark. It's a much mellower evening than yesterday and still very warm. There's a smell to warm air that seems exotic in the dark. I see shooting stars. There's much less traffic noise but machine-gun fire and explosions from the Battle Area instead. No bats but the exquisitely sweet song of a chirruping toad somewhere down by my feet. There are small frogs and toads everywhere at the moment. Smell of blackberries, plums, crab apples. Happy.

10 SEPT | WAXING CRESCENT

The moon has set, but I didn't see it. 21:00 Moz is gorging on blackberries: she picks ripe ones by smell, so darkness is no problem to her. Crickets still sing. 2 muntjac are barking at each other. The tawny owl hoots. A tractor is ploughing in the distance. Jets. I can hear crab apples falling. I can smell crab apples and damp. Duck overhead, 2 shooting stars.

Moz goes after one of the muntjac. She doesn't bark; I just know she's gone. Without seeing, I know what's happening. I saw Luther yesterday and he had a bandaged hand. He told me he had been into one of his sheds to look for a bucket of screws which he knew was on the floor. He hadn't bothered turning the generator on, he had just gone into the black, windowless maze of sheds and kicked where he thought the bucket was. As he kicked around in the dark he said he knew something was coming at him. Luckily he put his hands up over his face because something hit him and ripped the back of one of his hands. It was a tawny owl. Luther's homemade tin sheds are perfect nesting and roosting sites for owls: they are dark and dry, don't have solid doors and, most importantly of all, are very rarely visited. We had a long conversation about how he knew something was coming towards him without being able to see or hear it. He said he FELT a whooshing.

Without vision, hearing and smell are obviously heightened, but Luther knew something was coming at him and I knew Moz had gone, without either of us hearing anything. I wonder if there is some sense of movement that we get through our skin?

13 SEPT | FIRST QUARTER

22:00 Half moon hardly rising above the horizon today. It rose at about 15:00 in the SE and has just set in the SW. I've been finding large mottled moths in the house every morning. Looked them up, they are called Old Ladies. These are much darker than the example in my moth book, but it's definitely them. I put a particularly ravishing one outside in the woodshed this morning. A purple sheen on its wings and huge, vivid, red purple eyes. If only I could see in ultra violet. If only I could see the world through a moth's enormous eyes.

17 SEPT | WAXING GIBBOUS
Moon low in the S. 20:45. Warm. I'm standing by the big pool: an otter is squeaking as it travels down the stream. It's coming closer and closer. I hear its splashes and bow waves as it swims past. It doesn't see me and I don't see it. A never-ending procession of huge tractors and trailers on the main road. Orange lights flashing. V noisy. Pink moon directly behind a long cloud almost the same width as it. All pink gold fuzziness with just its shockingly luminous bottom hanging out. A gold shooting star in front of Cassiopeia.

18 SEPT
23:30 Waxing gibbous. Blustery clouds moving fast from the W. I'm in Cumbria. Damn Mellbreak, it's so in my face and blocking the view to the S. It's difficult to get a hang of the sky, so I go up into the field behind the house to see if the moon is visible from there. It's not, but I know where it is. The fast-moving clouds are disappearing behind Mellbreak and the leading edge of the ones lower in the sky are turning pink as they approach the fell. On certain nights Mellbreak is the perfect representation of the Greek myth of Endymion, the beautiful youth who lies in a perpetual sleep in a cave on Mount Latmus, and who Selene, the moon goddess, loves and visits every night.

When the moon is out of sight behind Mellbreak, her presence is always clear, her radiance haloing the sharply jagged outline of the fell. When the moon is behind Mellbreak, shining her light into the crevices of its southern flank, its northern face is the blackest black in Cumbria.

After a while, it's not just the leading edge of clouds but the whole billowing cloud that is illuminated pink, gold, cream of tomato soup colour as it approaches the edge of the fell. It's like an open door spilling light onto the street and the people arriving are momentarily lit up before entering. The unbearably slow intensification of the light, the movement of the clouds, the solid nearly vertical line of Mellbreak, then the moment arrives, the tension breaks, and the moon steps out into the sky. Saffron and so low. It's right at the bottom of the mountain. Saturn is just visible at the top. The moon is so low that the shadow of the wall is about five times its height.

20 SEPT
In Cumbria. A front has blown in, the air is thick with mist and low cloud. The full moon invisible, there is no bright point where it should be. The mist and cloud has soaked up the moonlight and disperses it in illuminated arches and wedges. Gradations of silver and grey.

It is very light. Grasmoor is reduced to a faintly visible South Downs-sized bump topped with a great silver dome at least three times its actual height. Above my head, dark wedges emanate from Mellbreak. The invisible has been made visible and the visible invisible. Lustred droplets in the saturated air catch in my hair, on my jumper.

Suggestions for names for the September full moon: rowan berries, hawthorn berries, sloes, gorse flower, pink feet, ticks – the dogs are covered in them. I have never seen so many. Picked 20 off Moz yesterday.

21 SEPT

In Cumbria. Huge plates of dark grey cloud the size of whole valleys slide in at sunset. The waning gibbous moon is visible fleetingly at 22:00, making rainbow edges of the cloud above the cloud it briefly slid out of, before being covered up by another cloud.

The sound of pink feet all night.

24 SEPT

First thing in the morning. The fells are invisible. I watch a large skein of pink feet fly low over the house into the murk towards Grasmoor. As they come out of the shelter of Mellbreak and over the lake, their formation breaks apart. They seem to turn right and head off in a ragged tangle down the valley. 20 minutes later they reappear in smaller groups flying back into the wind. Just above the house they are hardly moving, calling to each other, powerful, steady wing beats inching them through the misty currents of air. Usually the geese fly straight over the top of the fells on their way south in the autumn. You usually catch the occasional call from somewhere high, high above. But this cloud has forced them down and they have been channelled into a trap of zero visibility, up drafts, down drafts and back eddies between high fells. All day more geese arrive, are funnelled into the valley and have to turn round. The sodden maelstrom of sky is full of pink feet, at tree height, going in all directions. Pulse after pulse urgently heading south, many to the Norfolk coast. I'm heading home in a couple of days, too. I'm hoping low cloud won't be an issue but there is panic buying of fuel to contend with as well as roadworks on the Doncaster bypass. The full moon and the autumn equinox sun invisible to me behind layers of thick cloud isn't such a tragedy. I don't need to be staring at a clear sky to observe the pivotal moments of the year. Here now, the even distribution of day length over the whole planet is being marked by the urgent and perilous migration of geese.

26 SEPT | WANING GIBBOUS

Back in Norfolk. Clear sky. 07:00 The moon is high in the W. I'm surprised by the size of the missing chunk. It looks a strange shape this morning

– almost as if I've drawn it! Its top edge seems flat. I'm wondering if it's to do with the position of the maria? Am I seeing the moon from a slightly different angle? Has it wobbled?

The large, usually darker, splodges on the moon are called maria (singular: mare) because they were once thought to be seas. They are actually flat plains but they were given evocative names by the 17th-century astronomers who were obviously as seduced by the beautiful moon as the rest of us.

- Mare Vaporum – Sea of Vapours
- Oceanus Procellarum – Ocean of Storms
- Mare Marginis – Sea of the Edge
- Mare Crisium – Sea of Crisis
- Mare Spumans – Foaming Sea
- The rather less poetic Mare Smythii – Smyth's Sea
- Mare Tranquillitatis – Sea of Tranquility, where Apollo 11 landed on 20 July 1969; Neil Armstrong and Buzz Aldrin took their giant leap for mankind there…briefly shattering the tranquility, I'd guess

18:15 Sun setting due E. Probably a blessing for my eyesight not to have had a sunset to squint at for days.

Moon rises 21:00 very far to the NE. Fat and orange red. Tawny owl. Stars. The moon is sitting on the roof, bigger and brighter than the lit window from the house, but the same orange. Mixing a colour for the moon you'd certainly use the illuminated window orange, but there's more red in the moon. Darker, punchier, stronger and more luminous. Heavenly light. Sunlight through stained glass is probably the only way to properly replicate the dark luminosity.

27 SEPT | WANING GIBBOUS

Sunset W 18:40. Oh I love drawing fast. 4 pieces of paper stuck to the board. 4 or 5 minutes a drawing.

- Sun on left of birch tree, obliterating the right-hand side of the tree.
- Sun behind birch tree, obliterating most of the tree.
- Sun to right of birch tree, obliterating the left-hand side of the tree.
- Sun eating the tops of the trees on the horizon slightly further to the right of the birch tree.

Light spilling this way and that. I'm just inside the gate to the big field and the buffalo are in here somewhere, so I'm also scanning for them occasionally. Feeling of elation, something to do with speed and colour and surprise. Pheasants making a noise from the wood as they go up to roost. Shots from beyond the lake.

22:00 Very still, completely clear. Drawing the rising moon. Really enjoying it. The Pleiades are sparkling like ice. Oh no, I hear the gamekeeper's little buggy coming down the road. He works on moonlit nights, too. I can't concentrate until I hear him driving away. No shots tonight, thankfully.

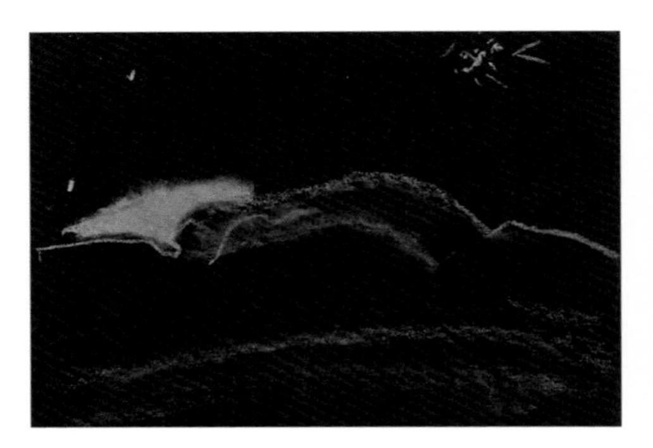

I can see Slinky looking hangdog outside my studio in the bright moonlight. I decide to finish up and take him in. I turn my head slightly and see a big bright blueish spot of light falling through the sky. It's much, much bigger than a star. Much brighter. Directly under it is a triangle of paler light with 2 stronger beams projecting to the corners. It has a curved base. There is a dark band of sky under that and then a more diaphanous cloud underneath. It's falling very slowly down towards the N horizon. It disappears behind the trees, over the village. I'm appalled and I'm excited. Dread and something I can only describe as awe are running amok in every nerve fibre, every cell in my body. WHAT THE HELL WAS THAT? Too big and weird for a star. Definitely not a shooting star – I've never seen one like that. Nothing like a satellite or an aeroplane. No torch or laser ever looked so beautiful or big or graceful. I only have two plausible suggestions – a UFO or an angel. Yes, angel seems totally reasonable right now. I don't know much on the subject but I've long suspected they're not the saccharine young sweethearts depicted on Christmas cards. I've just seen something that is by turns horrifying and beguiling. It seemed elemental. The UFO theory is too frightening to contemplate while I'm out here. I'm totally freaked out, my heart is thumping, and I decide that I'm never going out in the dark again. Am I going to sleep tonight? Will anyone believe me? The dogs are unconcerned, both now happily hunting for voles.

28 SEPT | THIRD QUARTER

Half moon. Everyone was asleep when I got in last night and #UFOinukrightnow didn't come up with anything. As I brushed my teeth the world seemed normal, but how could it? I'd just seen a falling angel. The calm, rational, pragmatic part of my brain was having to damp down the sparks of anxiety which kept igniting.

The news this morning didn't mention angels or UFOs either. So thank goodness for Twitter. My timeline was full of photographs of last night's angel. Apparently it was 'Centaur upper stage from the Landsat 9 launch performing its deorbit burn over the British coast', i.e. the rocket that had just launched the Landsat 9

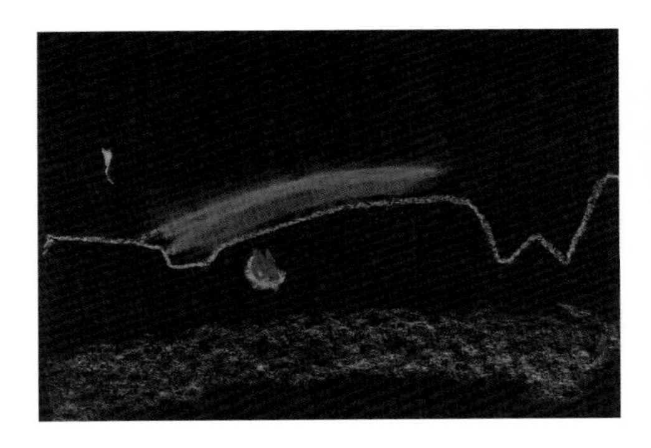

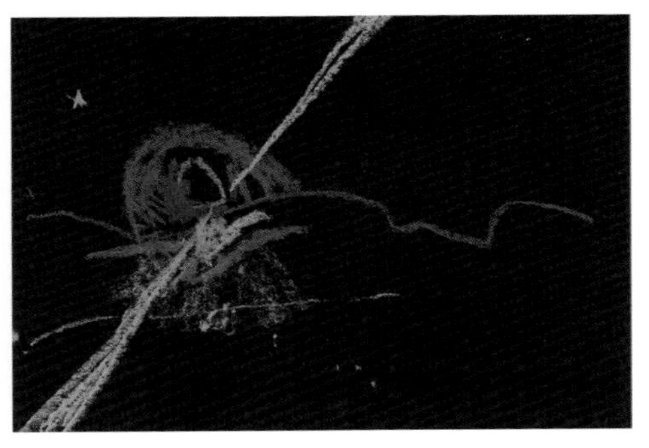

opposite & above 27 September 2021

satellite into space re-entering Earth's atmosphere. The launch of Landsat 9 had been on the news yesterday. Thank goodness for that explanation. I had spent last night wondering if I was sliding into one of those American docudramas about people who swear they've seen an alien spaceship and spend the rest of their life shut in the attic drawing little green men. Actually, as I write this, I think my life IS quite like that.

22:00 Rain. Crunch of snails underfoot as I take the dogs out. It feels extra dark tonight as I've just been looking at my computer screen. With unusable eyes I have to try and keep to the path by feeling my way with my feet and shins. Slow deliberate steps and for some reason I'm keeping my feet quite rigid and flat; perhaps it's more secure when I know there are one or two holes about. My shins are leading the way: it's with them that I can feel the taller grass on the edge of the little trampled track I'm trying to walk along. It's a strange, slow-motion way to walk. My connection with the ground is the most important thing, and so it feels like all my senses are in my feet and lower legs. I can't walk in a straight line, though. I keep hitting the edges. It's weird because I know this path is fairly straight. On the last length, my eyes are adjusting and 2 giant puffballs flare out of the darkness to light my way.

29 SEPT | THIRD QUARTER

06:30 The rain of last night has blown through and it's a cloudless morning but noticeably colder. The moon is unbelievably high in the S. Almost overhead. Way higher than the sun ever gets. I thought the sun and the moon and the planets all moved along the ecliptic. The moon seems to be the only one that goes this high. Always the quarter moon. It's in the sky for 17 hours again today. But this is the third quarter moon as opposed to when I noticed it in Feb and March – that was the first quarter?

I've been thinking about the rocket. Does 'performing its deorbit burn' mean it was destroyed as it re-entered the atmosphere? Or were its thrusters, or whatever they're called, doing a burn (like they do on takeoff) on its way back to California to be re-used? If it

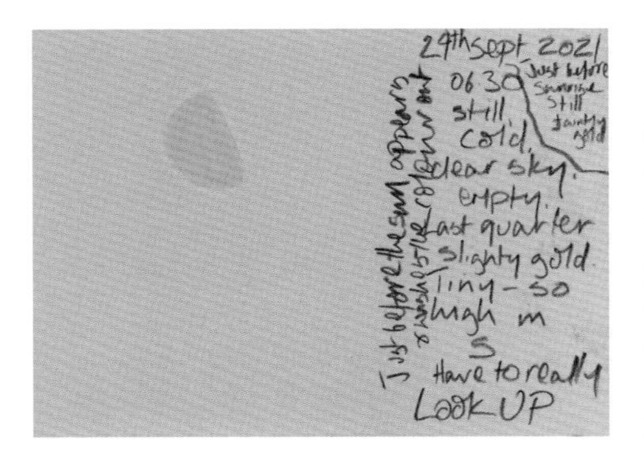

was destroyed in the atmosphere on the way back, how did it get through on the way out?

30 SEPT | WANING CRESCENT

There are a lot of mallards on the lake. I think someone must be feeding them. Their quacking sounds like laughing. One duck quacks and the whole lake erupts into hysterics. I told Ruby a joke, she didn't laugh, but the ducks did.

22:30 Very still and warm. Starry. The Pleiades are visible in the E. The ducks continue to laugh. Can't help but laugh with them.

3 OCT | WANING CRESCENT

19:15 Walking past my studio, I notice how low Arcturus is: it's setting in the W, just behind the wood. A few steps further on I wonder what the bright light, way above Arcturus, is. Is it Jupiter? No, Jupiter's over there in the SE. Venus? No, it's not visible at the moment. A star? What star? It's brighter than Jupiter. I'm wracking my brains, I almost turn round to get my planisphere from the studio, when I notice that it's moving. I guess it must be the ISS. I watch it travel out of the W and into the zenith. I look away and take one or two more steps and when I look back up…it's vanished. I scan the sky but can't see anything moving. As I walk the dogs round the field thinking about it, I realise that I was seeing it because it was in sunlight. Although the sun set here an hour ago, the curve of the Earth and the high orbit of the ISS means that I saw it reflecting the sunlight it was moving through. It disappeared when it flew into night.

Really understanding that I'm living on a spinning sphere. Really feeling the spin.

4 OCT | WANING CRESCENT

Nearly new moon, breeze. 20:30 So much noise tonight. I think John Daniels is harvesting his potatoes. Various machines scream and clonk and strain. Occasionally the whole S sky lights up. The end of the world seems to be happening on the other side of the valley.

5 OCT | WANING CRESCENT

21:00 Heavy rain followed by heavy showers all day. The wood is roaring like a wild sea tonight. Gusts of wind hit the poplars and a second later the birch outside my studio bends double. Every tree or group of trees makes a different noise. I love the hiss of the

above 6 October 2021

Scots pine. The grass joins in. All I can see in the dark is silhouettes thrashing about, but I can hear the cacophony and I can feel the movement of the air on my face. I saw my first flock of redwings today.

6 OCT | NEW MOON

Willow branches on the ground this morning after last night's wind. At 16:30ish the cloud cleared and the wind dropped. Walk to the far field to draw the sunset. I disturb teal and mallard on the lake. The sun is behind Mr Girling's wall of overgrown Christmas trees. Drawing fast. Colours very sunset-y, oranges, purples, yellows. I draw the colours and the clouds change in gaps in the trees. Like a kaleidoscope. I thought the horizon was invisible behind the trees until suddenly a distinct red line flares up in what had been a wall of black. After my last drawing and at the moment the sun sets, a mist begins to rise. A dark blue stain is creeping up the dark wall of trees at the bottom of the hill. It's the place that I'm sure the river used to run, years ago, before this landscape was so altered. It's much the lowest point and the mist is always thick here. Suddenly there is dew on the grass. This has all happened within minutes of sunlight disappearing.

A fieldfare flies over me with a pink underside and there is a slightly fishy damp-pond smell.

21:30 It's cloudless and misty. It's cold standing in mist. I've got very cold feet. Very cold ears and nose. The mist is rising and falling like breath. 2 rifle shots from the golf course.

8 OCT | WAXING CRESCENT

The moon should be visible after sunset tonight. I look for it all afternoon, holding my hand up to block the sun, but I don't see it. 16:20 Walking the dogs round the field, I glance up. As I look down I feel a familiar shape shock and I look back up – there's nothing there. And then I see it – the palest pucker in a turbulent ocean of cumulonimbus. A small white cloud to its left packs a heftier punch. I stop dead in my tracks, feeling quiet pleasure at locating the moon, feeling strangely conspiratorial. At a new moon and a full moon you don't need charts and tables to know where the moon will be. It's basic – you just need to know where the sun is.

10 OCT | WAXING CRESCENT
Wind in NW. 17:45 Can't bear being stuck in the valley bottom surrounded by trees. It's an autumn thing. So I walk up to the crossroads for a complete panorama. An amber sunset with long pink and purple clouds radiating S and E. Waxing crescent moon, missable and low in the SE in the colourful and busy sky. A long pink and grey cloud just above the moon evaporates. It's there one minute and the next time I look it has vanished. It's nowhere to be seen in a suddenly empty sky. It's lovely up here, but the traffic on the main road is distracting. Is that car turning left or right? Are the dogs standing in the middle of the road? Will that shiny 4x4 slow down? I can hear a steam train. A car pulls up and a lady jumps out to photograph the sunset. Beautiful purple blue pink in the E. Moon beginning to turn pink on its leading edge. By the time I'm back at my studio it's glowing in a pink gold sky above the lake. Every sort of bird is making a noise. There's mist down here and it's cold. 19:45 Moon red and very low. 22:50 The moon has set and all the beauties are beginning to gather in the E above the mist. There's Capella and Aldebaran, Taurus, the Hyades and the Pleiades.

11 OCT | WAXING CRESCENT
07:00 Sun low and watery. Dewy cobwebs in the field especially catching the light on the dark yarrow. Above, the apricot and silver glow of the sun is caught in a spider's web of contrails. I count about 18. From the track side of the field they seem to be centred on the sun. Fine brown red shadows along the sides furthest from the sun. Just like the cobwebs which are visible and then become invisible as I move, so the shadows on the contrails disappear as my position in relation to the sun changes. The world can change in one or two steps.

20:30 A fanfare of mother-of-pearl rises to a crescendo of pink and red to herald the red gold moon as she descends from the bottom of a blanket of blue cloud and appears to roll over the horizon on her back.

Friends and family now bring me news of the moon. Last night she was spotted descending Little Dodd. She was seen above the Roman Catholic cathedral and the other day was caught loitering at the bottom of York St. She is often to be seen trailing her silvery fingers across the Taf estuary. And I've heard that she sometimes sets over Scunthorpe.

12 OCT | WAXING CRESCENT
20:30 Very cold. Moon looks much bigger than last night. It seems to have jumped 2 or 3 sizes. It's weird when it does this – it isn't consistent. With huge chunks of the moon becoming visible each night, time seems to be galloping. It's rising in

the SE and setting in the SW, hardly managing to get above the tree line. At the last full moon it seemed to stay the same size for days, rising and setting in the same place. Time appeared to be at a standstill. I'm beginning to not care that I don't understand it and can't predict it. Perhaps it doesn't matter. It hasn't diminished the enjoyment and the wonder so far, has it?

Drawing the moon behind the poplars. It bursts forwards obliterating foliage and trunks, and what it can't obliterate by sheer force of light it manages to subdue with a vast pale blue halo. I look at the trees in the morning and I can't believe the moon managed to punch a hole through that tangle. I can only see the occasional glimpse of daylight.

Taurus rising in the E. Aldebaran looking really orange.

13 OCT | FIRST QUARTER

Warm, cloudy. 19:30 First quarter moon not visible but lovely mother-of-pearl smears and streaks in the sky. Owl hoots and a woodcock gives me a fright bursting out of the grass from under my descending foot. The potato harvest is still audible. All day Eric has been carting potatoes to the farmyard. The road is littered with beautiful, white, unblemished potatoes. I filled my hat with them and made a Spanish omelette.

15 OCT | WAXING GIBBOUS

18:00 At exactly the time when the sun sets, the moon begins to glow. Weakly and pink at first but quickly becoming gold. The sky must become much darker for this to happen. But because I'm viewing it against trees which are also getting darker the sky appears to stay light.

Clive's red Herefords, at the far end of the meadow, looked extraordinarily red as they caught the last of the sunlight this evening. Much later, as I looked across the stream, that end of the meadow was completely lost in mist. As I looked, someone turned up in a truck to check on them. Yellowy headlights and powerful white lamps on the top of the truck searching in the mist for invisible red cows. When the lights did eventually pick them up, they had been transformed into fuzzy nuggets of gold floating in mid-mist.

23:00 The moon very bright. Forming an obtuse triangle with Jupiter and Saturn low in the ESE. These pattens and shapes are really what sky watching is about. When you see a nice line-up or arrangement, you can comment on it. A conjunction (when two objects in the sky come very close to each other) is just an astronomer's name for the excitement of things lining up.

16 OCT | WAXING GIBBOUS

Moon rising just before sunset. 22:00 Silence, minding my own business,

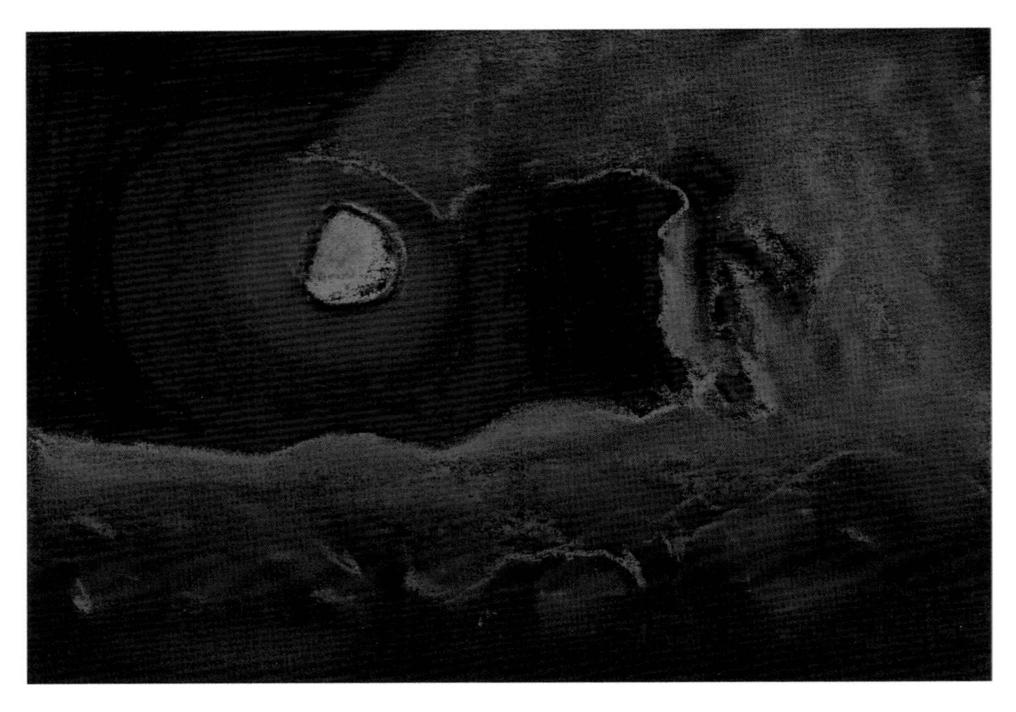

drawing blurry moonlight on shadowy thorns, when I'm briefly caught in a strong beam of light. It sweeps right, then back towards me. Then right again. Oh God, I'm about to be shot by a poacher. Where's Moz? The safety of darkness, the dusky magic of moonlight destroyed. I can't tell where the light came from. It could be a mile away. It could be on the other side of the lake. But I do know that all bets are off after dark. I know that risky shots are taken towards roads and footpaths, shots that wouldn't and definitely shouldn't be attempted in daylight. I'm in my own field, but all feeling of safety and isolation is gone. I head home.

above 17 October 2021

17 OCT | WAXING GIBBOUS

20:30 Very bright moon. On first going outside and looking up, the sky appears green blue, a sort of dark blue jade. But after a minute or two the greenness seems to fade. I wonder if this is something to do with the warm yellowy electric light in the house that my eyes have been tuned into. The moon is in a clear bit of sky, but an arc of altocumulus swirls round it and eventually in front of it. I now prefer drawing the moon in a cloudy sky. The off-balance rings, arcs and blotches of colour are always changing and always unexpected. The delicacy of moonlight exposes subtleties in the shapes of clouds which direct sunlight obscures. Heron. Moorhen.

18 OCT | WAXING GIBBOUS

We moved the buffalo onto the field next to the house yesterday. The stalks and seed heads are tall and dead, but underneath the grass is deliciously thick with late summer regrowth. The buffalo look magnificent in the pale dead grass, like they're on the African savannah or North American prairie. When they're lying down chewing the cud, they're almost invisible. It's only the outward thrust of their dead-grass-coloured horns that give them away amongst the vertical stems.

Thick cloud, rain and drizzle later. I walk into the big field at sunset under a lid of dark grey cloud only faintly pink in the W. The moon is obliterated. Unexpectedly the field is bathed in a mustardy glow. All the khaki greens of this morning are yellow or orange or red or brown now. The wet, dead ragwort stalks are the colour of sun-dried tomatoes. The sedge and the blackthorn are brassy. About a hundred crows are dancing over the rust-coloured poplars. Greylag geese. Light fading quickly, eyes unable to focus clearly. Wish I'd brought my sketch book with me.

19 OCT | DAY BEFORE FULL MOON

Very strong W wind. Even more overcast this evening than yesterday. Crows whirl round as if being stirred in a giant cauldron. I think about all those crow toes clinging on to bendy poplar branches all night in this wind. I try and draw the incredible mustard afterglow that I saw on the big field last night but it's not quite so strong today. I think the cloud is even thicker. I have a go anyway and there is a precise moment, in the 3 seconds or so between looking down at my paper and looking back up again, when all the colour just vanishes. No moon visible tonight. Rain.

20 OCT | **FULL MOON**

Afternoon – I'm looking at a rainbow and wondering why the moon's halo only occasionally has green in it and never violet. Moonlight is reflected sunlight. But maybe not the entire spectrum? In moon halos the red is always very prominent; in fact, it often goes beyond red to brown.

At 17:30 I set off to find a place to draw the full moon rise. It will rise at 18:04, N of E. This is a tricky direction: the wood is in the way. I end up down in the bottom of the big field looking up towards the dip in trees between the wood and Luther's yard. I disturb a flock of fieldfares in the blackthorn.

• Drawing 1 - The sun has set and everything up to the wood is in shade, the wood is glowing intensely gold, and the sky above is dark. Noisy geese. Crows.

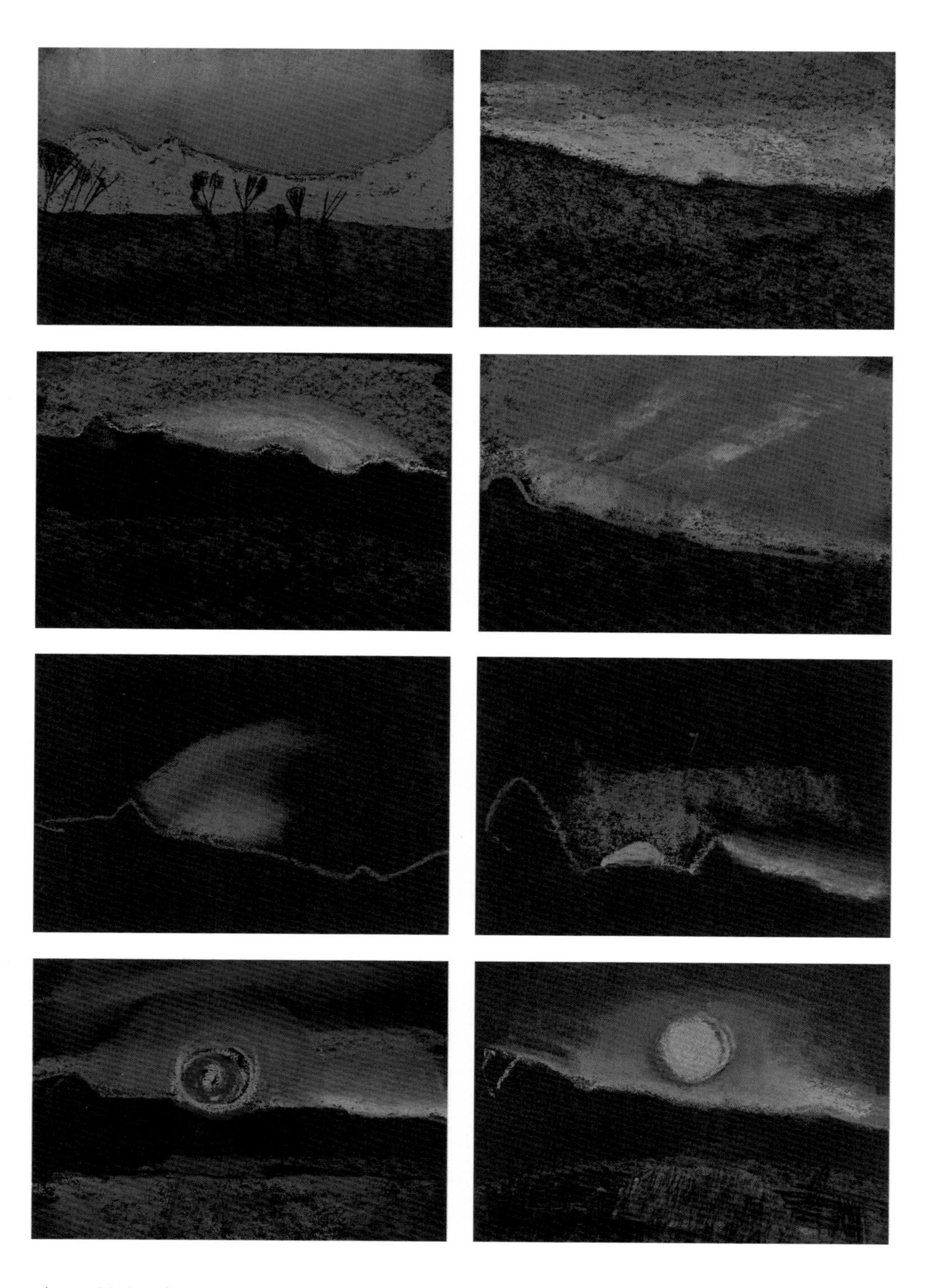

above 20 October 2021. Drawings 1–8, left to right, top to bottom

• Drawing 2 - The gold fades and the wood merges with everything else on the ground. The sky appears to brighten. The horizon line is the main thing. Ducks.

• Drawing 3 - A faint pink glow is discernible behind the wood. I can't see my pastels any more.

• Drawing 4 - The sky above the glow gets darker as the glow brightens. Owl.

• Drawing 5 - The glow is yellow, the glow is pink. The glow intensifies. The field and wood are black.

• Drawing 6 - The shockingly bright top of the moon is breaking cover over the top of the dark tree line. Clear-edged in the smear of Battenberg cake light around it.

• Drawing 7 - As it tops the wood, the field comes back into view.

• Drawing 8 - Higher up and I can see my pastels again.

Surely this is the best show on earth? Watching the moon rise is a cheap and sure way of inducing euphoria. My body is buzzing with endorphins or dopamine or whatever it is that makes me feel so pleasingly high.

22:30 A sheet of dark grey stratus is extinguishing the moon. It is a faint icy ghost when I walk out of the back door and has disappeared without trace by the time I get to my studio. Brutal, like witnessing a death by smothering.

The need to draw everything – I mean the need to draw the 360-degree dome of the sky and the spin of the Earth – is probably best done by drawing in a series, either by time or by direction. Well, that seems to be the best attempt I can make of it.

Suggestions for names for the October full moon: woodcock, fieldfare, hip, haw, wind, potato.

21 OCT | DAY AFTER FULL MOON

18:17 Clear night. Much colder. Watch the moon rising from the road bridge. HUGE and behind a scribble of cloud. I hear golden plover.

23:30 I walk up the track later. Moon shadows, shadows in the dark, are something I'll never not be delighted by. Seeing them is a reminder of the astonishment of new things seen in childhood. I stand under a crab apple and look up – the moonlight catches the curve of every apple and they shine. It's difficult to tell the difference between crab apples and the stars visible through gaps in the foliage. Rather than colourless in moonlight, I think everything is looking green tonight. Crab apples, leaves, bracken, track. I'm thinking about dark chromium greens when I hear a faint noise from the field and see a large dark shape. The buffalo are floating/tiptoeing towards me in a tight group. They are hesitant, bunched up, they stop and sniff. They come nearer. They move away. They come back. Six hugely fat water buffalo move as one, almost silently, like a will-o'-the-wisp through moonlight and dead grass stalks. I stupidly say 'hello buffalo, it's only me', thinking my familiar voice will reassure them. It breaks the spell and they fall apart. Going in all directions, they begin to kick and dance and grunt, running towards the fence and stopping with only inches to spare. Accompanied by six, wild, ten-ton-Tessies, Moz and I hurry back down the track, hoping like hell they don't come through the fence.

22 OCT | WANING GIBBOUS
06:45 Sirius bright in the S. The moon, high in SW, is exactly as it was last night, but with the lower right edge slightly less defined. Moonlight in the wallow that the buffalo have made on the big field. A heron has taken to following them round the field by the house and 2 magpies are pecking about and sitting on their backs.

There's a burnt-out car in the entrance to the golf course this morning. I wonder what the poor buffalo made of that last night?

23 OCT | WANING GIBBOUS
07:00 The moon has been rising at about the same time every night but is setting an hour later. It's climbing higher, too. By about 10 degrees a night. This morning another big chunk has gone. It looks like the back of a pudding bowl haircut, one that finishes at ear level.

It just occurred to me that perhaps the moon is actually changing as I look at it. That the sun is retreating from the missing bit as I watch. I'd always just assumed that it was a different size every time it rose – as if some magical change had happened off stage, as if props and makeup had been busy arranging the new look. Now I think about it, that's obviously absurd. Perhaps if I lived in a place where there were days at a time when the moon didn't set, it would have been obvious that this is what happens. That it's a continuous change. There's something about the moon's rise – its drama, an entrance par excellence, a definite beginning – that leads you to believe its shape must be set (for that day or night, at least). Perhaps also it's the way we view it. A glimpse here, half an hour there, a picture in a book – it seems fixed.

19:10 The moon is a cochineal egg rising to the N of the oak tree that is my NE marker. It's almost free of a thick band of dark cloud. Just the bottom is obscured. The section of cloud silhouetted by its vivid light is sharply in focus; it appears to have a perfectly scalloped edge. I stand still for a long time. I remember standing here last October watching the dark red moon rise. This must be an October thing, although I know the moon rises in the NE at all times of the year. It's very starry. Crab apples crash to the ground occasionally and a muntjac moves around on the edge of the wood, directly behind me. I think I give it a terrible fright when I eventually move and I hear it bouncing off down the bank. It barks loudly when it's a safe distance from me. Cars heading N on the main road keep catching my eye.

24 OCT | WANING GIBBOUS

21:30 Really spooky night. Warm, windy, waning gibbous moon casting pearly light down behind dark clouds that resemble phantoms. It's the sort of sky from a horror movie.

2 shots from W.

The 2 men who stood motionless, side by side, apparently overseeing the removal of the burnt-out car, are stuck in my mind. They both had red beards. One man's was mid-chest length, paler in the middle, ending in an extreme, rigid point. A bit like a stiff upside-down candle flame. I really wouldn't like to bump into them in the dark. I make double-sure I've locked up. Although thinking about it now, they asked me what day the car was burnt and I answered that it was on the night of the full moon. That's a bit weird and witchy, isn't it? They were probably as freaked out by that answer as I was by them.

25 OCT | WANING GIBBOUS

Between twilight this morning and twilight this evening 6 ink caps have appeared on the edge of the track to the field. They are a foot high.

21:30 Marmalade-coloured moon rising in the NE behind the spectre-shaped Monterey pine. It's completely dark. A waning moon travels through the sky, missing edge first. Somehow this seems more heartrending than a waxing moon that travels with its full edge first. A waning moon is a shrinking, or some say dying, moon, which is obviously a sadder state of affairs than a growing waxing moon. I wonder if you showed a group of non-moon watchers two images, one of a waxing and one of a waning gibbous moon, both at the same size, which one they would say is sadder. Would they think there is any difference at all? Perhaps it's got more to do with when they are being viewed. The waning gibbous moon rises late, and lonely, in the dark. The waxing gibbous moon rises in daylight.

The moon has been extremely high in the sky recently. It's been above the horizon for 14 hours. It's another one of those hugely-high-for-a-hugely-long-time, quarter moon things. Happy just to notice these things with no intention of wracking my brains as to why.

28 OCT | WANING GIBBOUS

Moon low, growing faint in the NW at 14:00. Warm wind from SW. Sunny. Draw the setting moon in the big field and contemplate why I very rarely see the moon setting. Is it the geography here? Is it the moon's rhythms? Or perhaps it's my own rhythms?

Later I find 2 ticks on me. It seems that sitting still, even standing still, in the big field is just asking to be eaten.

29 OCT | WANING GIBBOUS
21:00 Clear sky. 2 owls call to each other, one in Luther's spruce trees and one in the trees in the middle of the field. I'm standing under a Scots pine looking E, at Aldebaran, hoping I might see a Taurid. A branch above me creaks and a third owl begins to screech right above my head. It's earsplitting. The three-way hullabaloo goes on for some time. I daren't move. I daren't put my fingers in my ears. One of Clive's cows joins in. Round and round, back and forth. Eventually the branch shakes and the shrieker heads off to Luther's spruce trees, screeching as it goes. The two in the spruce trees now make warble-y noises to each other. I see a faint gold shooting star. Rifle shot.

30 OCT | WANING GIBBOUS
16:00 The moon rose at midnight and has only just set. 18:00 After the moon, Venus is the brightest object in our sky and this evening it's really showing its brilliance. It's an un-ignorable rose gold freckle in the W just after sunset. The sky is still light: not even Jupiter, which is next on the brightness scale, can be seen. Venus is the Evening Star – it's also the Morning Star, and I have always been confused by this. Its orbit is closer to the sun than ours, meaning that when we look towards it we are always looking towards the sun. It's more often than not lost in the glare, just like a new moon. At the moment Venus is at greatest elongation, i.e. furthest from the sun, which means we get a good view of it. Sometimes that's in the early morning, just before sunrise, and sometimes in the evening, like now. We'll see it there for a few weeks and then it will become invisible against the sun. How you work out when it will become visible seems to be a dark art, but luckily you can find charts for everything on the internet and I've just looked at a table appropriately called 'Apparitions of Venus'. It seems there was a morning and an evening apparition last year, but this year there'll only be one evening apparition, next year there'll only be one morning apparition.
I guess you just have to make the most of Venus when it's visible. For the past few nights it's looked astonishingly pretty in warm peachy skies. Venus is actually in her girdle.

3 NOV | DAY BEFORE NEW MOON
21:00 No moon. Cold. Clear except for very bright aquamarine clouds on the N horizon. As I draw them they vanish and I notice the SW horizon is bright with them. They vanish, too. On every horizon banks of gorgeous blue clouds appear and disappear like magic. The sky remains clear. The world is very silent tonight. 23:50 Vega is setting in the NW. A sudden commotion of moorhens on the lake. A long-lasting, bright gold shooting star quite low, from E to N.

4 NOV | NEW MOON

03:00 A heron shrieks and wakes me up. I stick my head out of the window and look at the stars. Boring Aries is in the SW, it's a 2 star line with a dip to another star at the W end. Those books of illustrations of the stars always make me laugh and I pity the artists who have to make horses out of squares, kings out of houses, or a ram out of this dull line.

above 8 November 2021 (no diary entry)

7 NOV

In Cumbria. Waxing crescent moon. Couldn't find it! It set at 17:50 and the sky was clear, but it must have been behind something. Fell or trees? I'm always surprised by the orientation of things when I'm here. The Plough is scraping the top of the barn, as if it's in league with the cows.

Darkness hasn't brought peace to the fells tonight, though. Everywhere I look I see lights. Why trust your own eyes? Who needs glowing stones when you've got 200 lumens in your pocket at the flick of a switch. A bicycle club have turned their lumens on and are careering around Lanthwaite Wood like a swarm of glow worms. The lights are as bright as car headlights; they are reflected in the lake. I have a strong feeling that this is the wrong way to behave in a wood after dark.

9 NOV | WAXING CRESCENT

Home. 16:30 An all-guns-blazing sunset followed, in ascending order, by Venus, the crescent moon, Saturn and Jupiter. A perfect illustration of the invisible ecliptic. Nearly 2 years of moon/sun/stars and darkness drawing and it's dawning on me that it doesn't matter that I don't understand quite a lot. Not understanding is good; it keeps me looking. When I do recognise something, though, well that's worth enjoying. The colours of the autumn leaves stay visible long after everything else is lost to the gloom. The particular gold of the birch is valiantly keeping darkness at bay tonight.

10 NOV | WAXING CRESCENT

Waxing crescent invisible in enveloping mist. At dusk Luther's yard seems full of tawny owls. There are at least three. They sound like wheezy cats. They actually make a huge range of noises, which are mostly quite odd. They twit-twoo only occasionally.

Later, in the dark, the tops of the poplars merge into the sky. No sound except for the occasional drip. Then swan noises from the lake. Their calls are also quite varied. It's very warm. Smell of deep fat frying.

I wonder if drawing a familiar place in the dark is more difficult than drawing a place you've never seen before. I'm sure when I'm drawing here my brain is filling in gaps and indistinct areas with what it knows to be there in daylight. Am I actually drawing what my eyes are seeing or what my brain thinks it knows?

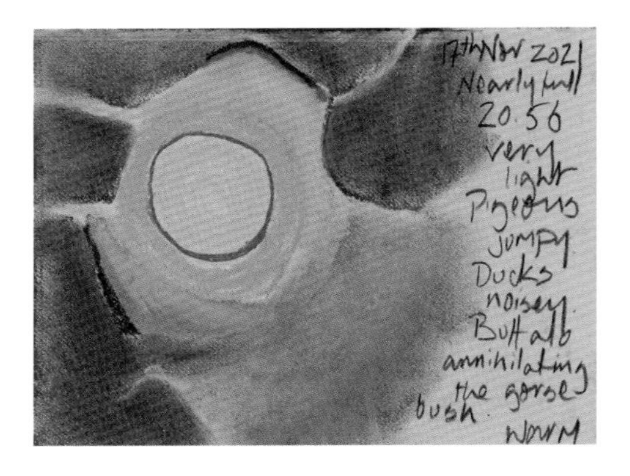

17 NOV | WAXING GIBBOUS

The moon has looked full for the past couple of nights. Full moon is not until the 19th. It's been rising at more or less the same time every afternoon but setting up to an hour later every morning. All this moonlight is making the world jumpy. Pigeons flap out of trees if they catch sight of me 100 yards away. Mallards and moorhens make a commotion all night and the swans are chasing each other around relentlessly. I can hear the buffalo annihilating the gorse bush in the middle of the field.

19 NOV | **FULL MOON**

15:50 The sugar beet is being harvested in the field on the other side of the crossroads. Started at 06:00 and still going. Lorries have also been carting sugar beet from the farm all day (they harvested theirs last week). Bang crash bang crash. The cacophony of machinery lumbering through mud or down the road, and the echo of tumbling sugar beet into empty trailers, is arable Norfolk's November song. Been a grey-brown day. Mud, water, noise. Don't expect to see the moon tonight. It'll rise NE. I stand by the river and draw in the rain anyway. Imagining where it's rising behind all the layers of cloud. It gets dark. 17:30 The cloud is breaking up by the time I get back to my studio with a cup of tea. 5 minutes later I look out of my studio door and a monstrously dark cloud is fleeing S, its serrated N edge pink and gold and turquoise. Suddenly the moon appears to race out from behind the cloud and stand still above the house.

No owls, no duck, no deer. Just farm machinery and lights.

Suggestions for names for the November full moon: sugar beet, flighty pigeons, teal.

21 NOV | WANING GIBBOUS

One day after full moon. 17:00 I had to go food shopping this afternoon and the rising moon accompanied me. Framed behind trees to my left, hanging over the bluebell wood in front of me, illuminating Carleton Forehoe church and turning the E sky airforce blue. It left me where the street lights started and I had to search it out when I parked outside the shop. It was just one of the many orange blobs lighting the carpark. Jupiter, Saturn and Venus were small-fry compared to the red lights on the police mast above me.

19:00 I'm drawing the moon roll along the tops of the poplars. Pink is the colour of things here tonight. The mist hugging the thorns is a cold pink below. The moon has spun herself a warm pink cocoon of cloud above. The teal on the lakes are loud.

23 NOV | WANING GIBBOUS

Sunny day. Moon setting behind the wood 10:30. We moved the buffalo to the big field and Andrew has taken Big Horns and Laurel to the abattoir. Tried to draw the pale moon between birch trunks with the bright colours of the field maple hedge but felt desolate thinking about their last journey. Somehow drawing the setting moon seemed to make it worse. They will be dead by the time it next rises. Drawing this place, their place, today is unbearably sad. My heart isn't in my drawing and instead I get a strange urge to carve their images into a fence post. I'm not sure why. A sort of remembrance? A memorial. But when I think about it, I realise they have been carving this place with their horns, their tongues and their teeth, with their sheer bulk, for three years. That Big Horns and Laurel were here is written large – in the chomped laurel hedge, the ringbarked alders and the muddy wallows, the trashed gorse bushes…

The moon – it's just a cold-hearted marker of time.

We eat our supper in silence.

24 NOV | WANING GIBBOUS

15:50 Drawing sunset from the top of the big field. The golden bramble leaves in front of me vie with the sky for punchiest colour in the afterglow.

20:00 Drawing the heartbreaking rising moon. Big, golden and maimed. The missing part appears to have been torn from the flawless circle. The wound is still jagged. I draw in the dark, not looking at my paper. The gold moon casts silver light. I draw the light, the moon, the shapes of trees, and the endlessly folding and re-folding reflection in the stream.

25 NOV | WANING GIBBOUS

21:00 Mist. The orange moon is the only colour in darkness and swirling mist. As it rises above the mist, it silhouettes the zig-zag tops of one or two pine trees. A little bit higher and it begins to stain the mist pink. In front of me, around me, behind me, in my lungs, on my eyeballs, the wet air is pink. Higher up, the moon's top edge nudges into cloud which is

above Sunset, 2 December 2021

transformed into pink wings. A water rail screams.

26 NOV | WANING GIBBOUS

07:15 The fog/mist is pink in the E. Not the moon's doing but the rising sun. I look straight at the sun and feel like Dorothy when Toto pulls the curtain back to reveal the real wizard. I can see the disc of the sun and it's so small! Naked. Behind layers of fog and mist it appears smaller than the moon was last night. Its blinding glare has been doused and the sun is nothing more than a puny grey circle above John's plantation this morning.

And then I look up SW...and there is the moon, a pale not-quite-disc in blue sky – still out of reach of the mist, where it's been since I saw it rise last night.

> The sun and the moon appear, to us earthlings, the same size. The sun is 400 times as big as the moon but it's 400 times further away. A full solar eclipse is when the moon is directly between us and the sun. It fits snugly, exactly over the sun. What an incredible coincidence. I've come to realise, viewpoint is everything.

1 DEC | WANING CRESCENT
Heavy cold showers. Very dark. 18:00 Was confused by hearing rain against the kitchen windows but going out of the back door I found a clear sky and Jupiter shining brightly. Silver and gold shooting star out of the N, heading W. 19:00 Fox is barking from Luther's yard.

Having so much darkness means I get several goes at being outside, looking. I can go in and get warm, I can cook or wash up, answer emails or have a bath, and then go outside again. I don't even have to get up very early. I'm wallowing in a surfeit of darkness, and every time I put my coat on and open the back door I get a little thrill of excitement because I never know what I'll see.

2 DEC | WANING CRESCENT
Cold N wind. Clear day. Venus, Saturn and Jupiter very bright early on. Later thin cloud takes the shine off the stars and eventually covers everything. The ground is wet and the buffalo pats slippery.

6 DEC | WAXING CRESCENT
It's been grey for days. Cloud so thick it hardly gets light during the day. I wonder what I thought was so great about drawing at night as I trudge round the field with the dogs in the dark at 17:30. I seem to step and slip in every buffalo pat in the field. What's happened to your seeing-in-the-dark superpower, eh, Tor? Today I actually hate the dark.

8 DEC | WAXING CRESCENT
I had a blood test today and told the doctor I'd been looking at the moon and the stars; he told me he'd always wanted to be an astronaut.

9 DEC | WAXING CRESCENT
The sky cleared as the sun set and for a couple of hours there was a very clear view of Venus, Saturn, Jupiter and the waxing crescent moon. I insisted Frank come and have a look. He grumbled. 'I have seen the moon before, you know.' When he saw the line-up, though, he stayed outside, he even said 'wow', and he patiently listened to me name the planets and try and explain the ecliptic. When we went back inside, he said, 'That was actually quite interesting.'

A bit later I'm sure that the moon had moved, in relation to Jupiter. Can this be true? Is it moving at a different speed to the stars and planets?

A lot of satellites tonight before it clouded over.

10 DEC | WAXING CRESCENT
15:45 The nearly half moon is high in the S and the sun has set. The flimsy daytime moon transitions to glowing orb – from milk to molten lava via silver and rose and marigold – not in a smooth succession and not equally all over. One second it's sort of silvery, the next moment the top edge is

definitely warm and rosy. Then the back edge catches fire and is extinguished. A bubble of pink explodes at the bottom. Venus is visible, but it's only a pinprick of light.

17:45 Venus is huge, an unmissably bright blob on the W horizon now.

21:45 Red gold shooting star over my studio. The moon is darkening and now low over Luther's yard. It's very endearing, the way it seems to be lying on its back. Although this can't be the case as I'm sure a waxing moon is facing the direction it's going. It's actually plummeting face first towards the horizon.

Heron shrieks.

11 DEC | FIRST QUARTER

Quarter moon invisible behind thick cloud. Cold. The sky is the blue grey of lavender leaves. Heavy rain on and off. An owl hoots from a long way away. The weir thunders, but the wood shed is warm and quiet and smells floral. I think it's the ash logs. 23:00 2 long-lasting red gold shooting stars in the E.

14 DEC | WAXING GIBBOUS

Thick cloud, warm. Moon appears to be racing through the sky behind the clouds but there is only the occasional puff of air down here. As I walk up the track at 20:30 I'm caught in a sudden shower of leaves from an oak tree. None of the others are doing it. Is it just a tiny gust of wind catching only this tree? Or an individual tree shiver of letting go? Is my presence related? Are the birds who chose this tree as a good, warm roosting place 6 hours ago slowly being exposed to the elements as the night progresses?

15 DEC | WAXING GIBBOUS

16:00 Clouds cleared as the sun set. It's very warm and a group of gnats descend onto the pieces of white paper stuck to my board. In the E everything is pink – the vegetation and the girdle of Venus (although Venus isn't in her girdle at the moment, she's in the W). As the light diminishes, the oak trees get redder. Brighter briefly than the moon behind them. Even without the brick-coloured leaves on the oaks, this half hour after sunset always glows red. Someone pointed out that 'gloaming' is from the Old English *glōwan*, meaning to glow. It's a sort of pulse, or a throb maybe, that slowly intensifies until it's almost unbearable and then it just vanishes, and it's night. Drawing it always ends in failure. It's an uncatchable thing.

Without cloud it stays light for longer. The crows are being laidback about going to bed, but the teal scurry about as usual. I draw until I can't see my chalks and go inside knowing that the moon will be up until 04:00 and the forecast is for clear skies so I can come out again later.

Later the moon is so high there are hardly any shadows. I can see which colours I'm putting on the paper.

above 15 December 2021

The bright moonlight is green and is blotting out all but the twinkliest stars. A crow is calling from the wood. A moorhen is calling from the lake, and an owl is calling from E and then S and then N.

Rifle shots.

> Swedish dialectal and Danish *glo* also have the extended sense 'stare, gaze upon', which is found in Middle English.
> –Online Etymology Dictionary

18 DEC | DAY BEFORE FULL MOON

Thick fog. I've got a cold and I'm feeling grim. Moonlight through fog tonight is murky green blue. It feels mouldy, surgical and snotty.

19 DEC | **FULL MOON**

So much cloud it can't be seen. Nor can its position be guessed at. In the two years that I have been drawing the full moon rise, only one or two haven't been visible. That's pretty good statistics. I don't feel like drawing at the moment anyway, I'm feeble and thick in the head tonight.

Suggestions for names for the December full moon: mud, fog, rain, oak leaves, silage, darkness.

20 DEC | WANING GIBBOUS

At long last the clouds have cleared. The moon is 99.3% illuminated, so high and bright I can comfortably see to pick

the last of the kale. I'm not going out drawing, though. I'm going to bed. This glut of night feels luxurious. My cup runs over with winter darkness and I'm going to squander it on sleep.

22 DEC | WANING GIBBOUS
After the longest night, the prettiest morning. Freezing. Everything frosty. 07:00 From my S-facing bedroom window I can watch a little section of horizon smoulder, catch light and flash fry the whole sky vermillion.

07:45 The moon setting in the NW. It looks green behind bands of pink cirrus.

16:30 A multi-blue twilight. Silence except for traffic noise. As I walk round

above 29 December 2021

the top of the big field I disturb some crows roosting in an oak next to the track. They silently fly out of the tree and then begin to call. Suddenly there are crows calling from everywhere, the noise gets louder and harsher, and the air above me fills with crows. I am being called out. I am a troublemaker, a disturber of sleep, a disrupter of the normal order of things. I feel my intrusion keenly.

24 DEC | WANING GIBBOUS
The new expansion of the industrial estate on the old aerodrome (over a mile away) lights the NW sky very brightly. The end of the wood is silhouetted in front of an LED sky. A similar electric blue glow has recently appeared on the SE horizon. There's a particularly bright conglomeration of lights from the Hingham direction. The night sky is literally being gobbled up by light pollution in front of my eyes. Perhaps I'm the last person who'll be able to see the Milky Way from this field?

29 DEC | WANING CRESCENT
15:30 Sideways NE drizzle for days, then this afternoon the wind changed direction and hot air from the SW blew in and broke it up. The first pale streak of yellowy blue dazzled. The first colour I'd seen for days. In twilight I drew the brim-full new wallow that the buffalo have made. Outrageous colour moving round the sky.

21:00 I wandered around in the field looking at stars. Naming constellations. Then as the cloud began to thicken, the connections in constellations weakened and I couldn't tell which star was which.

Everywhere is so muddy.

The weir and the sluice are thundering.

4 JAN | WAXING CRESCENT
16:30 First sight of waxing crescent moon. I'm at the bottom of the big field and it's low, it's behind the poplar stems, in the orangey green afterglow. 17:00 I'm looking at it from an upstairs window. This is the perfect midwinter crescent moon viewing spot. It hangs in the sky just above the horizon at the end of the lake; it's almost framed by trees. Jupiter is above it, further to the S. This is the epitome of a moon view, the composition is good, and the colours are balanced. It's dreamy. I hang out of the window, drawing, listening to the owls call and the last of the crows as they finally go silent. But I like my drawing of the new moon behind giant obstructing poplar trunks best.

5 JAN | WAXING CRESCENT
17:00 The moon is very near Jupiter this evening. Still a lovely sight from the upstairs window, but a less good composition. A military jet, which I can hear and can identify by a red flashing light, enters stage right, loops round the infant moon, and exits back the way it came.

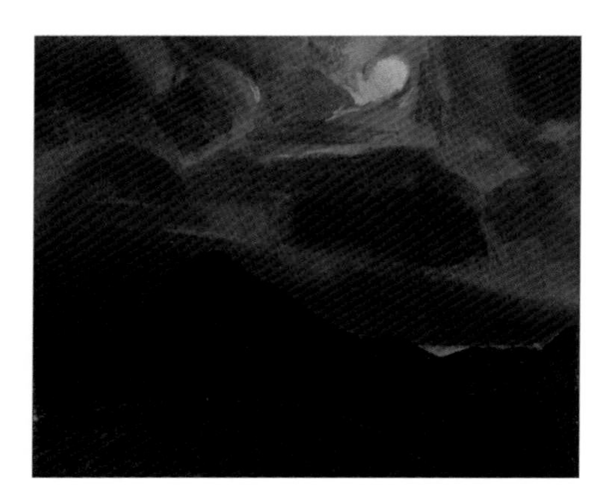

above & opposite 13 January 2022, Cumbria

Nearly two years of moon, sun and star watching and I may not have got to grips with the finer mathematical details (doubt I ever will), but I know what happens here. I have fully understood my horizons. I know where things happen, I know which places to go to in order to see things happen. I can feel the spin of the Earth.

9 JAN

Half moon. 16:45 Thick cloud. Very cold. In Cumbria, down by the lake. Watching the twilights perform their purple sorcery. Sound of waves. Ducks overhead. Two lights approach from the Mellbreak shore. A light comes on halfway up Grasmoor. There's a light on the top of Robinson. The walkers pass me without a word. Did they even see me? The light on Grasmoor is descending. There are three lights on the top of Robinson now. The invisible half moon makes tonight light and I walk back to the house without a torch. I can easily see where I'm going. I'm beginning to wonder if the point of a torch is to be seen rather than to see by?

11 JAN

Waxing gibbous in Cumbria. Mainly clear, sparkling and cold. See the chalky moon at 13:00 from the top of Gale Fell. It rose about an hour ago. Its pale reflection is stark in a heart-shaped peaty pool. 16:00 Low fell is yellow and the sky around it is yellower.

13 JAN

Waxing gibbous in Cumbria. Clouds. 17:15 Down by the lake again, in a pink and grey twilight. The moon is behind a stubborn, dark cloud above Grasmoor. On my way back up the hill the lights suddenly come on, or that's what it feels like. The cloud has cleared and there's the moon. Still behind cloud and spectacularly smeary and silvery pink. A clear bit of sky to its left astounds me with its turquoiseness. But only for a minute before the clouds reconfigure and things dim again. Slinky is completely invisible, but his shadow is walking next to me.

17 JAN | **FULL MOON**

Home. 15:20 Back in John's meadow. Two years since I first came here to draw the January full moon rise. Everything about that first attempt was a fluke. A fluke that I chose January

(I think January's is probably the gentlest of full moon rises). Lucky that it was a clear afternoon. And a one in a million chance that I was looking in exactly the right direction. My stars really were aligned that day. The stage was set and I innocently, incompetently, stumbled onto it. Two years on and I'm completely immersed in it but only slightly more competent. Drawing in the dark is still difficult, drawing a moving object in moving light levels is incredibly challenging. I still don't really understand the subtlety of the moon's position or timings, but perhaps chasing the heavenly being around my horizons has revealed something surprising, something much closer to home. It has taught me about the place I already know.

A pared-back box of colours sits on the ground next to me. Completely different from two years ago. No greens, just pinks and yellows and greys. Small bits of paper on a board.

15:40 Before the sun has set, blue yellow sky, a large half mint imperial above the pink bushes and trees. Not directly above the stream – just to the S.

15:50 The whole pale O is above the bushes and trees now. Only the very furthest are still bathed in the last golden pink sunlight. Warm pink is beginning to spread up from the horizon. Hasn't reached the moon yet.

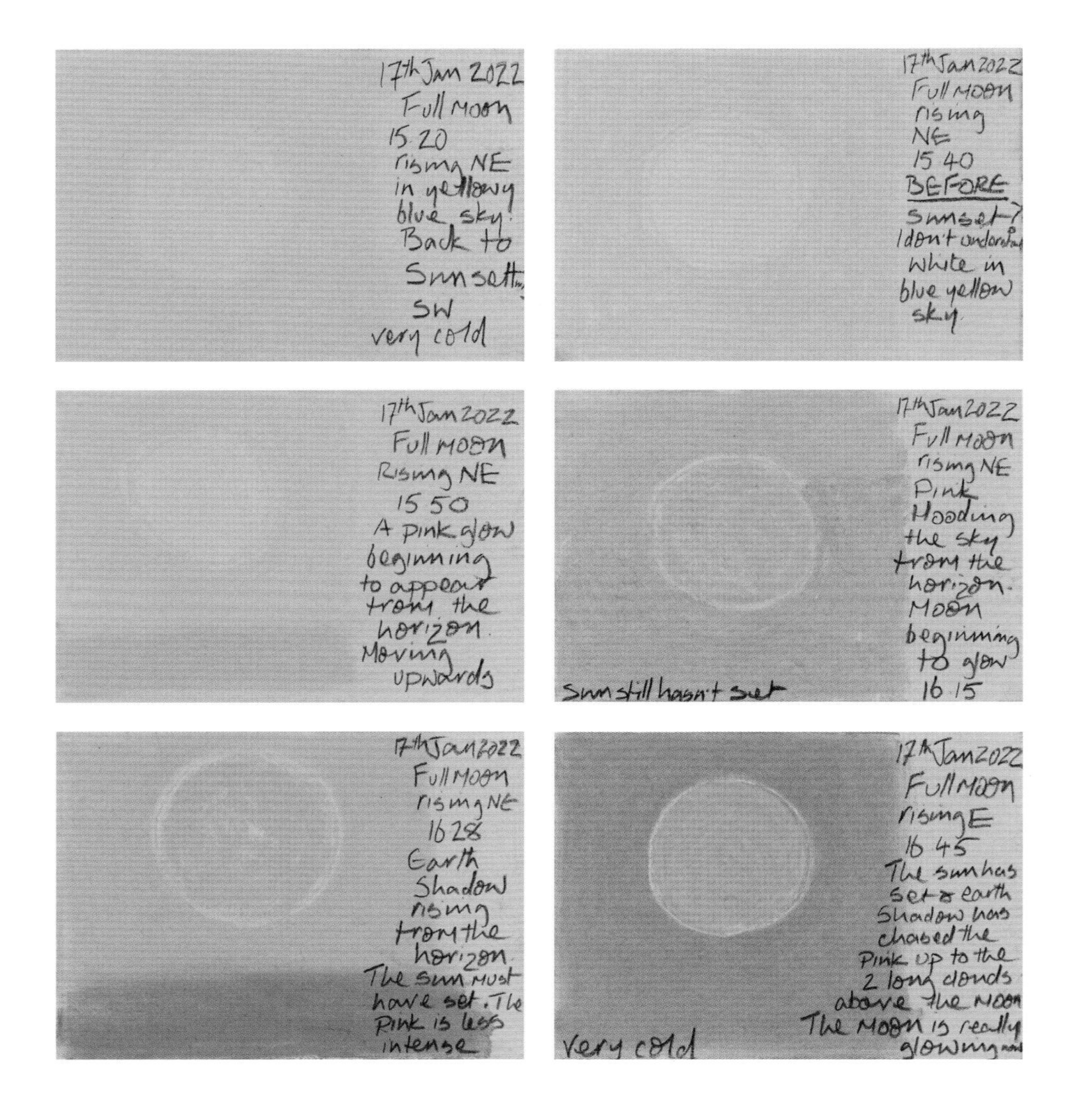

16:15 The sun still hasn't set. Pink getting more intense from the horizon and flooding up. It has overtaken the moon, which is beginning to glow.

16:28 Earth shadow climbing from the horizon. The sun must have set, the pink is less intense. The moon is gold. The moon is reflected in the cattle drink.

16:45 Earth shadow has chased the pink away. It only persists in two long clouds directly above the moon. The moon is orange. It is noticeably cold now.

My drawings are almost as bad as two years ago. New colours and a lot of practice haven't really changed anything. Slightly disappointed. Euphoric, too, though. Nothing beats going out to welcome a full moon rise.

18 JAN | WANING GIBBOUS
06:30 Frosty, last night's moon is above the wood in the NE. It has a dirty yellow glow around it. The wood is a reddish russet colour and the field maple hedge is alizarin; this year's long-growth stems stand out, they are pale, coated in rime. A robin sings. Ducks scurry overhead.

07:00 The moon is visible behind the wood, which has lost its reddish colour. The hedge is now scarlet and last year's growth is invisible. A layer of thick mist can be seen behind the hedge, between its tangled stems. A robin, a mistle thrush and a blackbird.

Enough. I find I'm repeating myself, standing in the same place at the same time of year, observing the same things. It's deeply satisfying – to be recognising the spin of the moon and the Earth through time and space. But I'm all out of colour words. Metaphors are running dry. Now, as the moon begins to wane, it's probably a perfect time to end all this recording. It's become exhausting – this always-on-high-alert looking. Almost exactly two years since I first purposefully sought out the moon seems as good a time as any to excuse myself from the close vigil I've been keeping on the ceaselessly turning circles. The Earth round the sun. The heavens round the Earth. The moon round the Earth. Me round my horizons. My two years of wonder are committed to paper. The next however many needn't be so rigorously documented. The moon took me out into space as far as I could see, and she showed me wonders and magic, but mainly she showed me here – this place – my place on Earth.

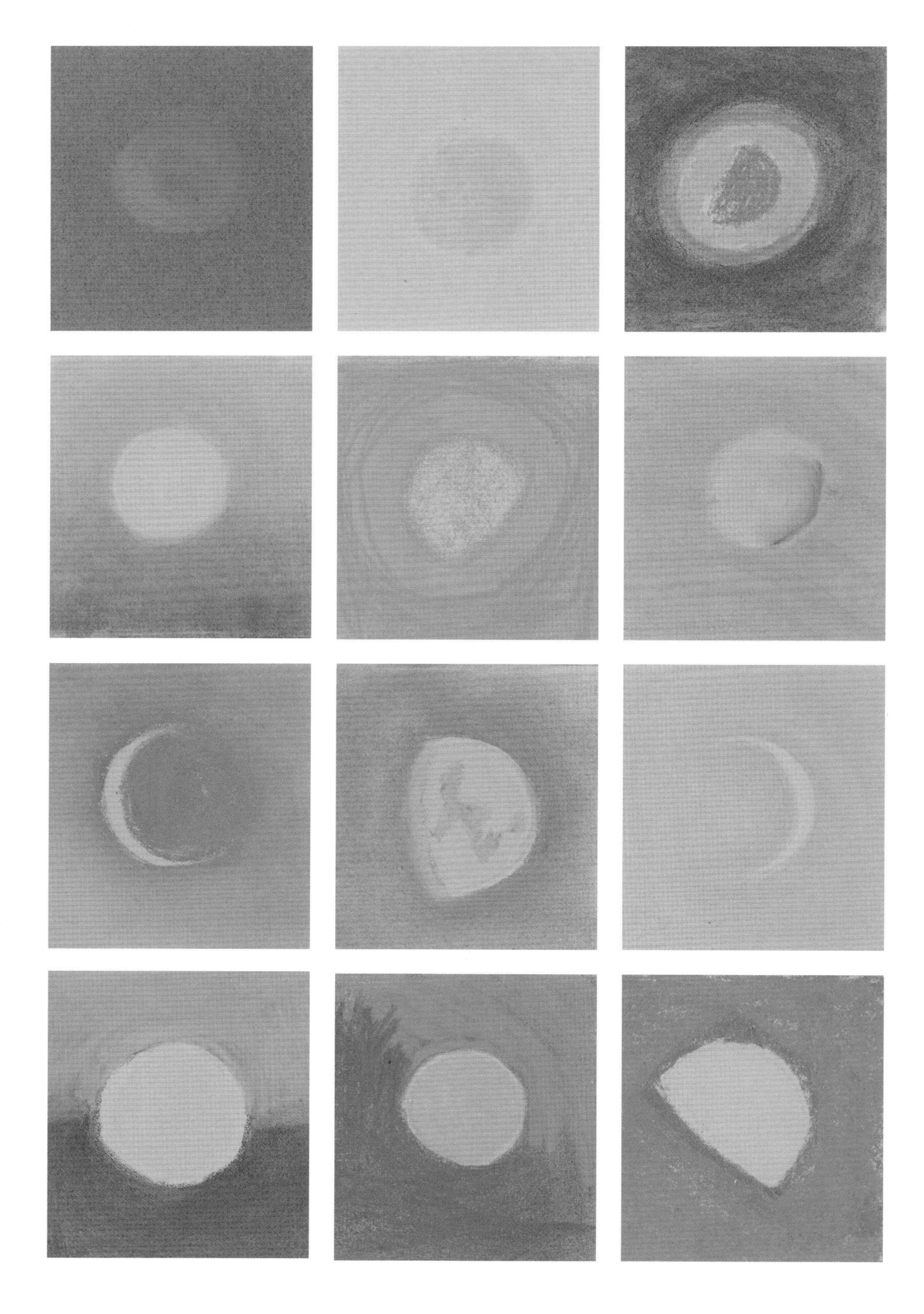

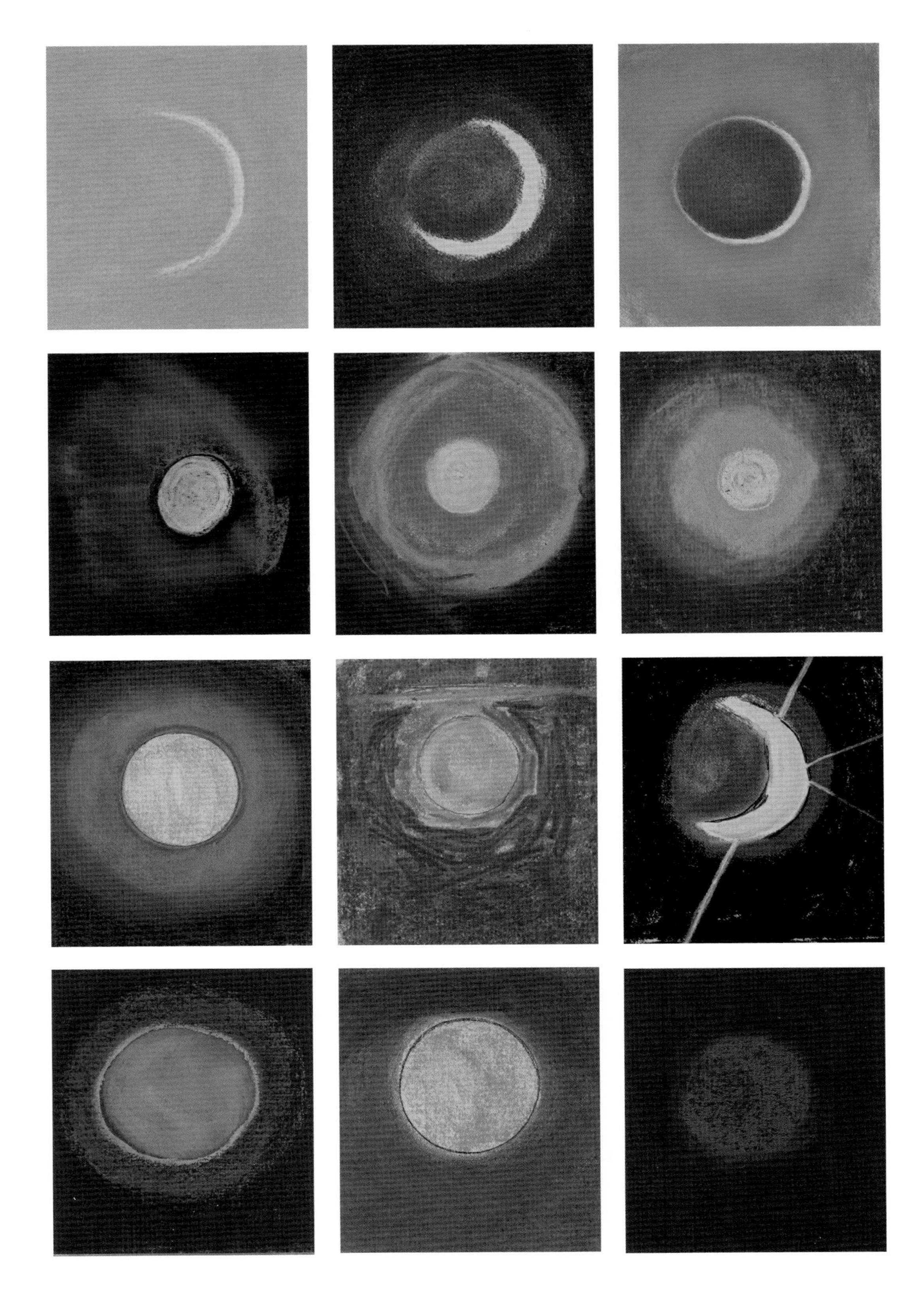

AFTERWORD

THE SCIENCE BEHIND THE MOVEMENT OF THE MOON

Dr Dan Self

Have you ever noticed how the moon is there one night, then a couple of nights later nowhere to be seen? Most of us are aware that it takes a month for it to go round Earth, so it can't have gone far. This sudden disappearance is more likely to happen in winter. Why? Well, it takes quite a bit of unravelling to visualise the moon's orbital path.

THE MOON'S ORBIT AROUND EARTH

The moon orbits at a more-or-less steady speed of 1 kilometre per second. It takes 29½ days to go through all its phases, waxing through crescent, quarter and gibbous to full, and waning back to new moon when it passes the sun, making it invisible.

The moon's orbit is more-or-less circular around Earth. I hear a lot about 'supermoons' these days, referring to the moon being closer, but the measured distance varies between 226 and 252 thousand miles, which doesn't really explain why the moon looks bigger sometimes. More likely, this is 'The Moon Illusion', an effect of how we perceive the sky near the horizon.

The moon's orbit is also more-or-less 'in the plane' of the solar system; it is only 5° off. Imagine the planets' orbits around the sun as looking down at track grooves on a vinyl LP and you are not far wrong. Viewing the record disk from a side perspective puts you within the plane, so the record is flat, and it looks like a straight line. Earth being embedded within the disk means the line goes completely around us and crosses our horizon at two points.

For visualisation purposes, it is useful to draw the entire circular path of the orbit out while the moon is obviously only at one point along it.

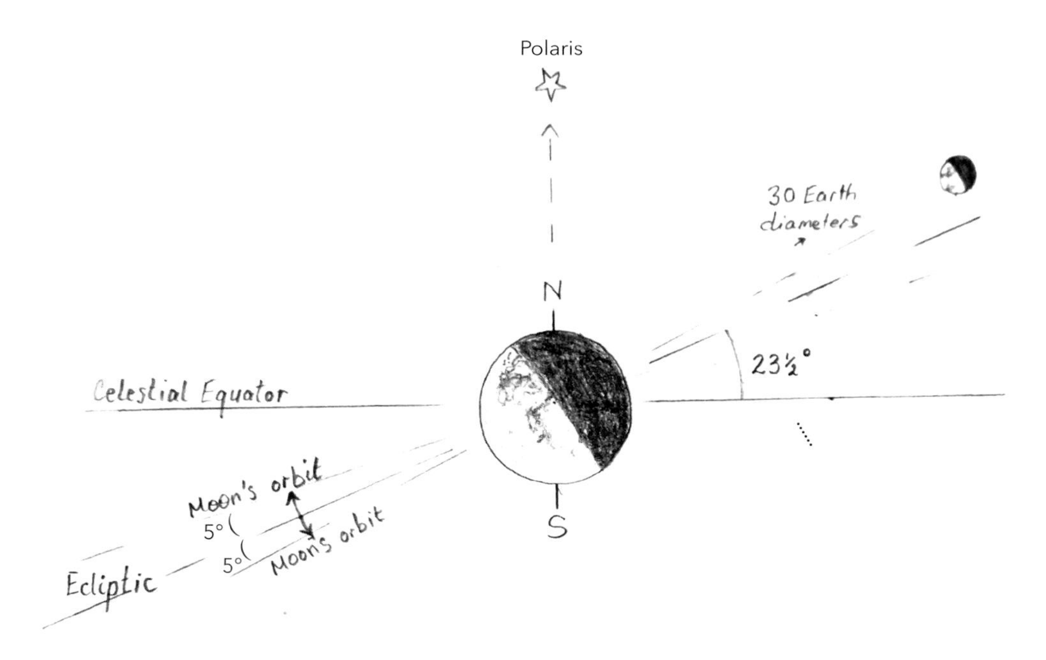

Diagram 1 Showing the orbital plane of the moon in relation to the Earth. The moon's orbit is tilted at 5° to the plane of the solar system (the ecliptic), and Earth's Equatorial plane is tilted at 23½° to this plane. The moon is around 30 Earth diameters away and about a quarter of the Earth's diameter.

THE CELESTIAL EQUATOR AND THE ECLIPTIC

Earth orbits the sun in a year within the plane of the solar system. From our vantage point on Earth, we see the sun, moon and planets all move along a narrow band that extends around the whole 360° of sky against a backdrop of star constellations. This celestial highway has been well known since ancient times. It was called the *Zodiac* and was responsible for horoscopes. A less astrological and more astronomical name for this line is the *Ecliptic* and it is plotted with respect to the distant star constellations. The ecliptic is the path the sun takes over the year through the stars, even though we can't of course see the stars behind the sun.

Earth's rotation is what causes day and night. The Earth rotates about an imaginary axis, which points northward and southward. If you extend this line out into the stars, by chance it happens to point to the star *Polaris*, in Ursa Minor, conveniently marking the North Pole. This star is the only one that is still with respect to our horizon here on the ground. All the others slowly rotate around it. You may have seen amazing photographs of star trails showing this effect.

Likewise, the Earth's south pole can be extended out into the stars, but alas there is no bright star anywhere near that direction in space. If you can keep visualising, perhaps you can extend all of Earth's Equator out into the sky. This forms a line on the sky called the *Celestial Equator*. This invisible line, like *Polaris*, does not move with respect to the ground. The three stars of Orion's Belt appear to move along it from left to right in the Northern Hemisphere, rising in the east, reaching their highest point in the south and setting in the west, 12 hours later.

The Celestial Equator is tilted at 23½° to the ecliptic. This causes the seasons. The ecliptic is sometimes north of and sometimes south of the Celestial Equator, i.e. half of the ecliptic is above and half below the Celestial Equator. Professional astronomers quantify this with Celestial Coordinates that they use to locate stars. But the misalignment between the solar system's orbital rotation (the ecliptic) and Earth's daily rotation leads to effects difficult to visualise.

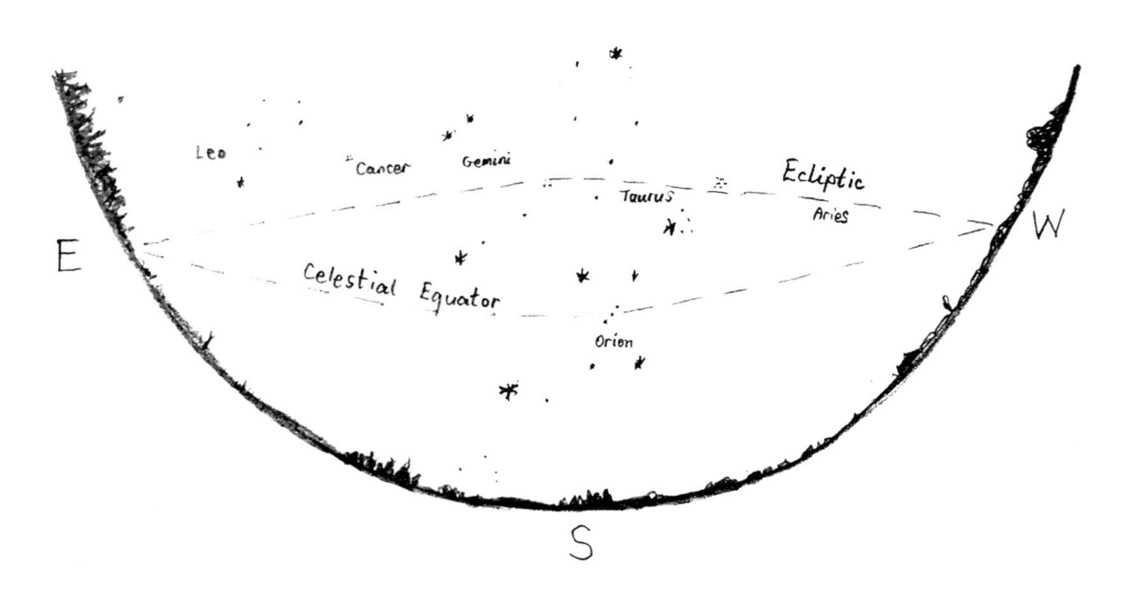

Diagram 2 Sketch of the stars behind the ecliptic featured as a dotted line through them. Constellations Cancer, Gemini and Taurus now pass above Orion. Orion's Belt stars lie on the Celestial Equator and are visible throughout winter.

As Earth endlessly turns, the sky looks like it moves around us. So, the imaginary ecliptic line seems to slink slowly back and forth throughout the night (and day). Also, the timing of this changes gradually throughout the passage of a year as we slowly move round the sun. The moon's orbit is almost on the ecliptic, so you can treat the moon's orbit and the ecliptic as essentially the same thing.

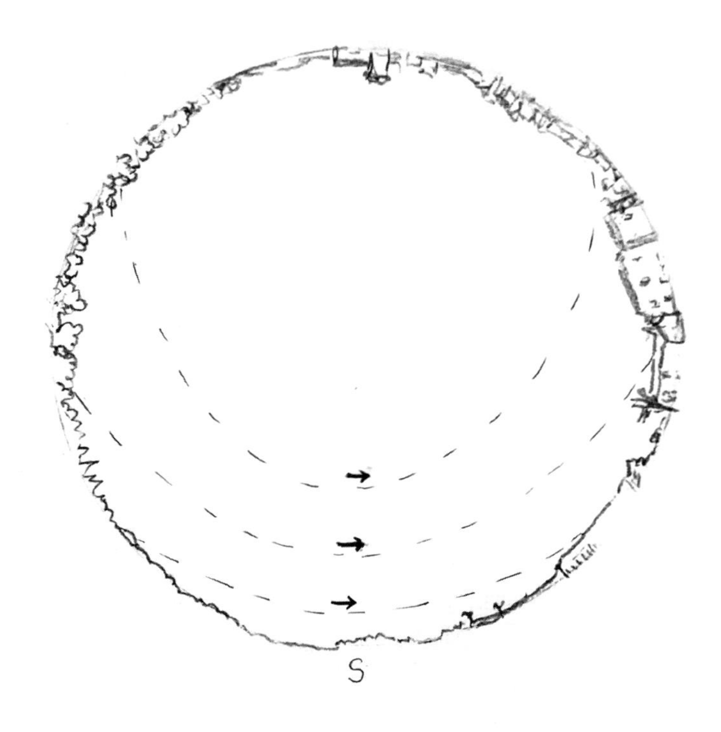

Diagram 3 Drawing of a wide-angle view of sky facing south from the UK, showing all the apparent paths of the sun, moon and planets across the sky as the Earth turns. The sun rises NE in summer and moves across the sky and sets in the NW. The path at equinoxes is E to W, and the path in winter is SE to SW.

From most places on the globe, the ecliptic can rise anywhere from our north-east to our south-east horizon. Likewise, it sets somewhere between north-west and south-west. Think about that midsummer sunrise in the north-east and that late sunset in the north-west and the high sun at noon. The sun is then at its most northerly part of the ecliptic, the name for which is the First point of Cancer. The southernmost point is called the First point of Capricorn, in the complete opposite direction to the former. Where the ecliptic crosses from south-to-north of the Celestial Equator is called the First point of Aries, and the north-to-south crossing point is Libra.

EXPLAINING MOONRISE TIMES ON CONSECUTIVE NIGHTS

It is these north-to-south crossing points where moonrise times become irregular. While the Celestial Equator maintains the same angle with the east and west horizons, the ecliptic and the moon's orbit sometimes make either a steep or a very shallow angle, as in the sketched diagrams on the following pages.

The **steepest angles** occur simultaneously at both eastern and western horizons at dawn in autumn, at midnight in winter and at dusk in spring. At this point the ecliptic is high in the south, near overhead.

Conversely, the **shallowest angles** occur at dawn in spring, at midnight in midsummer and at dusk in autumn. At this point the ecliptic is very low above the south horizon.

So around midnight in summer, the moon's orbit crosses the eastern horizon at a very shallow angle, especially in 2025, and especially if you are near the Arctic Circle. This has the effect of the last quarter moon rising on consecutive nights at almost the same time for a few nights but it will have moved to the left.

The same effect also occurs with the Harvest Moon at the autumn equinox. This point is near the First point of Aries, close to the constellation. To summarise, whenever the moon is at the First point of Aries, it is moving towards the northern leg of its orbit and objects on the Celestial Sphere that are more northerly appear higher from the Northern Hemisphere of Earth, where most of us live.

Diagram 4 Explaining the effect of moonrise times on consecutive nights when the angle of the moon's orbit is shallow to the horizon. Facing east, the last quarter moon in summer is near the First point of Aries. On the second night, it rises further to the north and only a few minutes later.

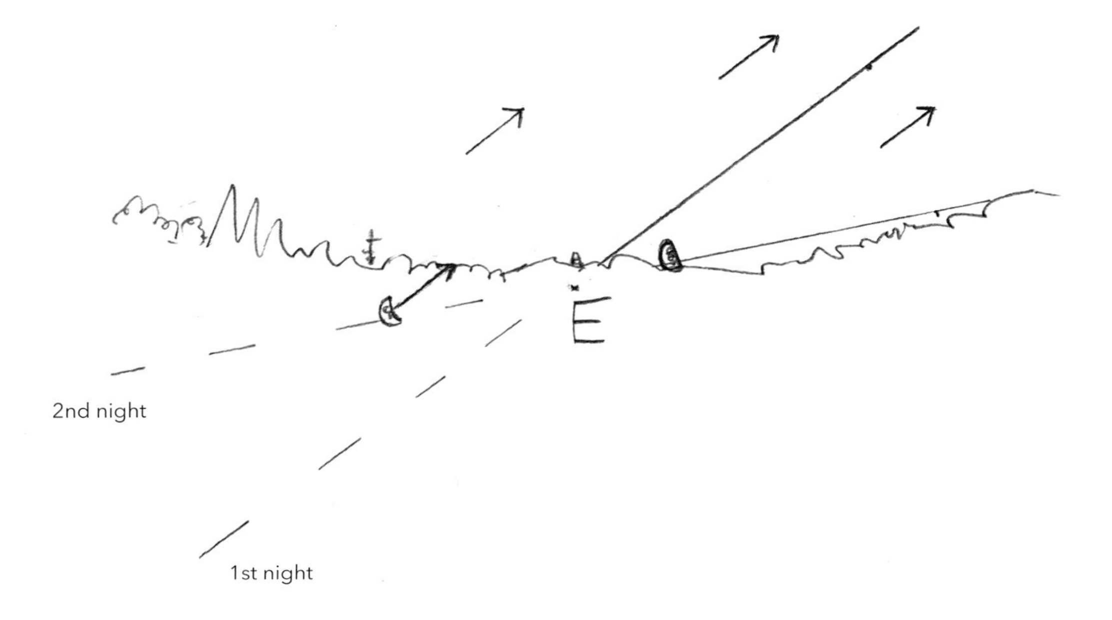

So, the opposite effect happens at the opposite time of year or opposite phase of the moon.

I will now refer to a moonrise where the moon's orbit and the ecliptic are at a steep angle to the eastern horizon. This occurs near midnight in winter with a rising last quarter moon or a full moon at dusk in March. The orbit at this point is in a more up-down direction when we are looking out at our eastern horizon. It is moving downward, so tomorrow it has further to rotate round until it rises above the horizon. It should rise well over an hour later, having moved to the right.

As for moonsets, the opposite is true. Going back to our autumn Harvest Moon: since a full moon is always opposite the sun, it rises as the sun sets, and, having passed across the sky, it will set around dawn. In autumn at dawn, the ecliptic is now at a steep angle to the horizon, and so moonset will be much later tomorrow as it is then more northerly and so has further to effectively move across the sky before reaching our horizon.

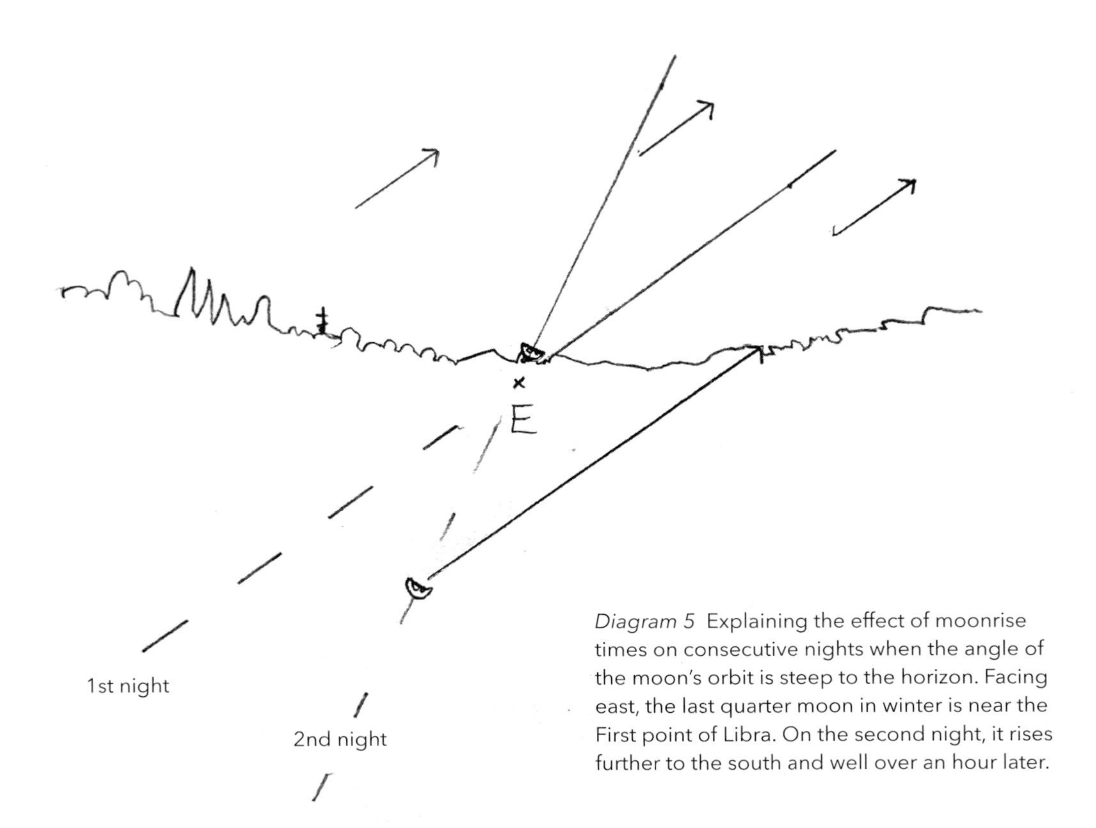

Diagram 5 Explaining the effect of moonrise times on consecutive nights when the angle of the moon's orbit is steep to the horizon. Facing east, the last quarter moon in winter is near the First point of Libra. On the second night, it rises further to the south and well over an hour later.

The most beautiful example of the moon's visibility is that in spring, when the new crescent moon appears much higher in the sky every night, lying on its back. This is because its orbit makes a steep angle with the horizon. In autumn, the waxing crescent moon is very low and positioned as if sitting on its point. All of this is explained by knowing what angle the ecliptic is.

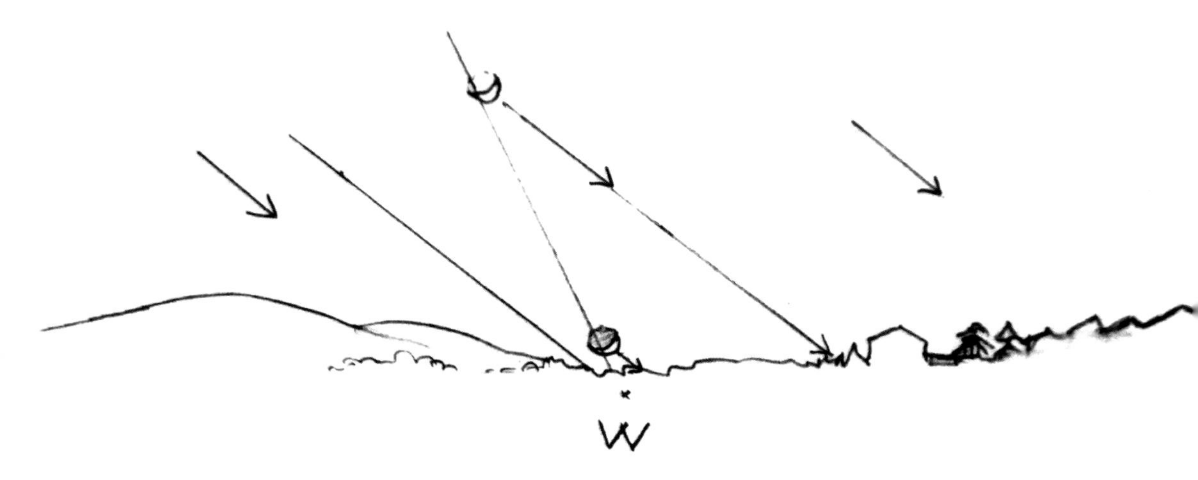

Diagram 6 The waxing crescent moon at dusk in spring. It rises rapidly over the course of a few nights.

We can also think of this as rapid lunar seasons that relate to our familiar solar seasons. The moon can be described as moving further north and south in Celestial Coordinates. When south, it rises later and sets earlier. North means it rises earlier and sets later, and thus it is just like comparing where the sun appears to be in winter to where it is in summer. However, as the moon only takes a month to go round us, it is like having the moon flitting between summer and winter positions every couple of weeks.

Of course, you could always look it up on an app, but that does little for your appreciation of the clockwork of the heavens. Spending years observing these patterns puts us in touch with our ancestors who instinctively understood it.